Special Report 242

CURBING GRIDLOCK

Peak-Period Fees To Relieve Traffic Congestion

VOLUME 1

Committee Report and Recommendations

Committee for Study on Urban Transportation Congestion Pricing

Transportation Research Board

Commission on Behavioral and Social Sciences and Education

NATIONAL RESEARCH COUNCIL

National Academy Press
Washington, D.C. 1994

Transportation Research Board Special Report 242

Volumes 1 and 2
ISSN 0360-859X
ISBN 0-309-05504-0 (Vol. 1)
ISBN 0-309-05505-9 (Vol. 2)
L.C. Catalog Card No. 94-16817

Subscriber Category
IA planning and administration

Transportation Research Board publications are available by ordering directly from TRB. They may also be obtained on a regular basis through organizational or individual affiliation with TRB; affiliates or library subscribers are eligible for substantial discounts. For further information, write to the Transportation Research Board, National Research Council, 2101 Constitution Avenue, N.W., Washington, D.C. 20418.

NOTICE: The project that is the subject of this report was approved by the Governing Board of the National Research Council, whose members are drawn from the councils of the National Academy of Sciences, the National Academy of Engineering, and the Institute of Medicine. The members of the committee responsible for the report were chosen for their special competencies and with regard for appropriate balance.

The report has been reviewed by a group other than the authors according to the procedures approved by the Report Review Committee, consisting of the members of the National Academy of Sciences, the National Academy of Engineering, and the Institute of Medicine.

The study was sponsored by the Federal Highway Administration and the Federal Transit Administration of the U.S. Department of Transportation.

Library of Congress Cataloging-in-Publication Data

National Research Council (U.S.). Transportation Research Board.
Curbing gridlock : peak-period fees to relieve traffic congestion / Transportation Research Board, Commission on Behavioral and Social Sciences and Education, National Research Council.
p. cm. — (Special report : 242)
Contents: v. 1. Committee report and recommendations — v. 2. Commissioned papers.
ISBN 0-309-05504-0 (v. 1). — ISBN 0-309-05505-9 (v. 2)
1. Congestion pricing—United States. 2. Traffic congestion—United States. 3. Transportation—Government policy—United States. I. National Research Council (U.S.). Transportation Research Board. II. Title. III. Series: Special report (National Research Council (U.S.). Transportation Research Board) : 242.
HE336.C66N37 1994
388.3′12—dc20
ISSN 0360-859X 94-16817
CIP

Cover design: Karen L. White
Cover photograph courtesy California Department of Transportation

Committee for Study on Urban Transportation Congestion Pricing

MARTIN WACHS, *Chairman*, University of California at Los Angeles
JOSE A. GOMEZ-IBANEZ, Harvard University
SUSAN E. HANSON, Clark University
STEVE HEMINGER, Metropolitan Transportation Commission, Oakland, California[a]
JOEL L. HOROWITZ, University of Iowa
LAURA A. JIBBEN, Regional Transportation Authority of Northeastern Illinois
RYUICHI KITAMURA, University of Kyoto, Japan
DANIEL L. MCFADDEN, University of California at Berkeley
MICHAEL S. MCGILL, Bay Area Economic Forum, San Francisco[b]
JOHN G. MILLIKEN, Venable, Baetjer, & Howard, McLean, Virginia[c]
PAUL E. PETERSON, Harvard University
MARK A. PISANO, Southern California Association of Governments
ROBERT C. REPETTO, World Resources Institute
WILLIAM T. ROACH, King County Metro, Seattle, Washington
KENNETH A. SMALL, University of California at Irvine

Liaison Representatives

JOHN W. ARCHER, American Automobile Association
JOHN T. BERG, Federal Highway Administration, U.S. Department of Transportation
DAVID J. HENSING, American Association of State Highway and Transportation Officials
KEVIN H. KRUKE, Highway Users Federation for Safety and Mobility
PAUL D. MCCARTHY, Ford Motor Company
TED SCOTT, American Trucking Associations
FRED WILLIAMS, Federal Transit Administration, U.S. Department of Transportation

National Research Council Staff

ROBERT E. SKINNER, JR., Director, Studies and Information Services, Transportation Research Board (TRB)
STEPHEN R. GODWIN, Study Director, TRB
SUSANNE A. STOIBER, Division on Social and Economic Studies, Commission on Behavioral and Social Sciences and Education
NANCY A. ACKERMAN, Director, Reports and Editorial Services, TRB
NAOMI C. KASSABIAN, Associate Editor, TRB

[a] Joined the committee in May 1993 to replace Michael McGill when he resigned.
[b] Resigned from the committee when appointed as Administrative Assistant to Senator Diane Feinstein of California.
[c] During the course of the study, Mr. Milliken was Secretary of Transportation, Commonwealth of Virginia.

Preface

In response to the growing interest in congestion pricing, the Federal Highway Administration and Federal Transit Administration requested that the Transportation Research Board and the Commission on Behavioral and Social Sciences and Education of the National Research Council conduct a study of this tool for congestion management. The study committee would

- Assess and synthesize available research and experience on congestion pricing,
- Commission papers on critical issues raised by congestion pricing to be presented at a national symposium, and
- Develop recommendations on the potential role of market pricing principles as a tool for congestion management, guidelines for the assessment of the impacts of congestion pricing experiments, and fruitful areas for further research, demonstration, or experimentation.

To conduct this study, the National Research Council appointed a committee with expertise in economics, political science, transportation planning, geography, transit and highway agency management, civil engineering, and government. The committee's deliberations were supplemented by liaison representatives from several groups concerned about the benefits and costs of congestion pricing. (Liaison representatives were invited to participate in the committee's discussion and review and comment on report drafts, but did not vote on recommendations.) After a review of the literature, and drawing from its expertise, the committee commissioned papers on a variety of topics. These papers were presented and discussed at a symposium held in Washington, D.C., in 1993. The papers, as revised by their authors after the symposium, are contained in Volume 2 of this Special Report. Volume 1 contains the committee's overview of the material contained in the commissioned papers, its conclusions, and recommendations regarding the potential of congestion pricing, the need for evaluation of early demonstrations, and other research needs.

Several of the members selected for this study committee and several of the authors of commissioned papers reside in California. There is good reason for this representation. Congestion pricing as a concept and proposal has been more extensively studied in California in recent years than elsewhere in the United States, and the only experimentation with congestion pricing moving forward at the time of this writing is occurring in California.

The report was reviewed by an independent group of reviewers in accordance with National Research Council report review procedures.

The report was performed under the overall supervision of Robert E. Skinner, Jr., Director, Studies and Information Services, Transportation Researach Board, and Susanne A. Stoiber, Director, Division on Social and Economic Studies, Commission on Behavioral and Social Sciences and Education. Stephen R. Godwin served as study director and, under the guidance of the committee, drafted the report.

The final report was edited and prepared for publication under the supervision of Nancy A. Ackerman, Director, Reports and Editorial Services, Transportation Research Board. Naomi Kassabian was the editor for the report. Frances E. Holland provided word processing support and, along with Marguerite E. Schneider, provided assistance in meeting logistics and committee correspondence.

Contents

VOLUME 1: Committee Report and Recommendations

Contents

VOLUME 2: Commissioned Papers

Executive Summary

Traffic congestion frustrates millions of American motorists every working day. The delay and wasted fuel from being stuck in traffic cost over $40 billion a year. Moreover, vehicles mired in traffic jams increase air pollution. With travel demand far outpacing the provision of highway capacity, there is little prospect that congestion will be eased simply by building new highways or transit systems. After billions have been spent on new highway and transit systems during the last two decades, it has become clear to many that America's metropolitan areas cannot build their way out of congestion.

Transportation policy is increasingly focused on managing the demand for transportation to alleviate adding capacity on new highways for use by solo drivers. Demand management measures include efforts to encourage ridesharing, telecommuting, walking, and use of transit. This shift in policy, combined with motorist frustration with congestion, environmental goals for cleaner air, and rapid advances in electronic toll collection, has renewed interest in an old idea—congestion pricing.

Congestion pricing would charge a premium to motorists who wish to drive during peak travel periods through strategies that could include tolls on roads or bridges, fees to enter congested areas, or changes in the structure of parking and transit pricing. When faced with a congestion fee, some motorists would decide to drive to another destination, at another time, or on another route; share a ride; switch to transit; or find some other option. These changes in driving patterns would reduce congestion and save time for motorists willing to pay the fee.

Congestion pricing could substantially reduce congestion, but it is and has been controversial. Motorists have become accustomed to paying for transportation through fuel, property, and sales taxes, and may greet congestion pricing with the same enthusiasm they have for new taxes generally. Groups representing motorists and commercial carriers at the local and national levels have long opposed tolls of any kind on roadways. Federal law restricts tolls on federal-aid highways and prohibits introducing new ones on Interstates (with the exception of congestion pricing pilot projects, which are discussed later). Americans have long supported gasoline and other user taxes as a means of financing and maintaining highways. Because user taxes charge for a service provided, such taxes are perceived to be fair—even though users with different incomes have different abilities to pay. Charging a premium for road use during peak periods, however, is perceived by many to be unfair to low-income motorists. Some may prefer to continue to use congestion delay as a "tax" that falls on rich and poor alike.

Because of interest in and controversy about congestion pricing, the Federal Highway Administration and the Federal Transit Administration requested that the National Research Council form a study committee to *(a)* assess the critical issues surrounding congestion pricing and *(b)* recommend the potential role of congestion pricing as a tool for congestion management, guidelines for evaluation, and fruitful areas for further research. The Research Council's Transportation Research Board and Commission on Behavioral and Social Sciences and Education formed a committee of 14 members. The committee commissioned 17 papers, which were discussed at a symposium held at the National Academy of Sciences in 1993. These papers, and one other commissioned after the symposium, are contained in Volume 2 of this Special Report. Volume 1 contains the committee's overview of the commissioned papers, its conclusions about the prospects for congestion pricing in the United States, and its recommendations.

BACKGROUND

A motorist who uses a freeway when it is severely congested pays the same money price to use that road as a motorist who drives outside the peak period. Whenever the price of using some valued good does not increase as demand increases, that good will be in short supply. Shortages will be acute if supply cannot be readily enhanced. This problem is typical of goods or services in industries with high capital or fixed costs.

Throughout the economy, premium prices are charged when the demand for some commodity or service exceeds the supply. This happens, for example, with telephone services, rooms at seasonal resorts, restaurant meals, and airline tickets. If businesses that have high fixed costs and variable demand provided capacity to serve all peak demand without raising their prices, their capacity would be grossly underutilized during off-peak periods and they would risk bankruptcy.

Businesses subject to peak demand and high fixed costs also try to balance demand and minimize capacity expansion by luring consumers to the off-peak period with lower prices. Telephone companies offer discounts for calls made at midnight. First-class restaurants offer lower prices for "early-bird" dinners. Resorts have lower rates during the off season. Air travelers willing to accept a less advantageous departure time or less direct routing can find cheaper flights. Consumers are quite accustomed to this kind of pricing, even if they do not always like it, and it keeps society from overconsuming the resources needed to satisfy its wants.

Americans are not accustomed to thinking about road use as a consumer good, but aside from the current lack of variable pricing, the analogy between use of a public good, like roads, and private goods and services, like telephones, airlines, restaurants, and resorts, is appropriate. They are all subject to demand that varies over time or space. They experience congestion when prices are not allowed to vary with demand. And in all these cases consumers have alternatives. Just as airline passengers can make choices about when to fly, which airline to select, or whether to fly at all, automobile users can vary the timing of trips, alter the route taken, carpool, use public transit, choose alternative destinations for some trips, or avoid certain trips altogether. These alternatives exist for many motorists even in the case of peak-period trips, because nearly one-half of these trips are not work related. If only a small fraction of the drivers in the peak shifted the timing of their trips, congestion would be greatly diminished.

In the absence of efficient pricing, motorists who drive on congested facilities are not required to pay for the delays they cause each other, and these delays are substantial. The wasted time and fuel are estimated to cost $40 billion annually. The environmental damage caused by the excess pollutants emitted during stop-and-go conditions drives the cost even higher.

CONCLUSIONS ABOUT CONGESTION PRICING

The essence of congestion pricing for transportation is to charge a fee that varies according with demand. Congestion pricing is usually thought of as charging a premium for the use of a particular road or bridge during peak travel periods. It also includes charging fees for entering a congested area or charging a premium for parking at certain times and places. Congestion pricing might well be part of a package of policies that would reinforce one another. That is, it could be combined with changes to the structure of prices for parking and transit services and the provision of alternative modes.

The committee's conclusions regarding implementing congestion pricing on U.S. roads and bridges are as follows.

Congestion pricing would cause some motorists to change their behavior.

Peak-period pricing has been successfully applied in long-distance telephone service and in many other sectors of the economy. Although specific changes in travel cannot be predicted with certainty, on the basis of estimates derived from past toll and parking increases, it appears that peak-period fees averaging $0.06 to $0.09/km ($0.10 to $0.15/mi), or $2.00 to $3.00 per daily round trip, would reduce total travel during the peak period by roughly 10 to 15 percent. This reduction would vary depending on the availability of alternate routes and modes, the amount of the charge, and the period during which a fee was charged. Most motorists are expected to continue to drive during the peak and pay the fee, but some would elect to share rides; change routes; drive at other times; change destinations; shift to transit, cycling, or walking; or forgo the trip.

Congestion pricing would result in a net benefit to society.

The shift in travel choices by a relatively small percentage of motorists would save time for the majority. These changes could be substantial: the average automobile commuter in a congested metropolitan area could save about 10 to 15 min per round trip. Such savings would cut the duration of the average commute trip by roughly 20 percent. Society as a whole would benefit. The aggregate value of the time savings would outweigh the added tolls. The hundreds of millions of hours that are currently wasted in traffic congestion each year could be spent in much more productive and pleasant ways. In addition, existing resources could be used more efficiently.

In order to compete in a global marketplace, all sectors of the U.S. economy are being pressed to become more productive. Without relying on pricing strategies to reduce peak-period congestion, society will be urged to finance new facilities costing billions, which will be raised from taxpayers, with the likelihood that these facilities alone would not reduce congestion. In contrast, if congestion pricing were adopted in all metropolitan areas, it would result in net savings on the order of $5 to $10 billion annually. Alternatively, to dampen demand in order to meet clean air goals, society would have to regulate travel behavior in ways less efficient than pricing. A pricing strategy would result in a more efficient use of resources and a more productive economy. By reducing the demand for capital while increasing output, the productivity gain for transportation would contribute to a more competitive U.S. economy.

Congestion pricing is technically feasible.

Congestion fees could be collected on existing toll roads in much the same way that users currently pay tolls. Alternatively, users could indicate that they had prepaid by displaying windshield stickers. Surcharges for parking in congested areas would not be complicated.

The concept of congestion pricing has been rejected in the past, in part because of concern that queueing at toll stations would itself contribute to congestion. Existing technologies in use on toll roads, however, allow tolls to be collected electronically without requiring motorists to stop. Emerging enhancements to this technology would also allow charging to take place without requiring individual records of the time and place traveled, thereby protecting motorists' privacy. Although these more sophisticated technologies are expensive, they would require 5 percent or less of the potential revenues in a regionwide application. Electronic pricing can be made readily compatible with the intelligent vehicle-highway system (IVHS) technologies that are to be phased in over the next decade or so.

Institutional issues are complex but can be resolved.

A variety of existing institutions could implement congestion pricing projects on individual routes or bridges. Congestion pricing on facilities throughout a region would be more difficult to administer because of jurisdictional issues and the lack of existing regional institutions to develop and implement such a plan. Metropolitan special districts, which are used for a variety of purposes in most states, could serve this role. Given the differences in existing laws and customs across states and metropolitan areas, institutional forms suited to each area would likely emerge if and when metropolitan areas are ready for a regional approach.

All income groups can come out ahead given an appropriate distribution of revenues.

Congestion pricing would generate substantial revenues. For example, by applying congestion pricing on all thoroughfares in the greater Los Angeles area during peak travel periods, over $3 billion in congestion fees would be raised annually. To ensure that all income groups benefit, some of this money would have to be returned in some way to road users.

On the basis of values of time derived from average wages, the highest-income motorists would value the time savings more than the toll. Whereas many middle-income motorists would pay the toll (if the alternatives were worse) and thereby save time, the time savings would not be valued as highly as the cost of the toll. In other words, before the revenues are distributed, the average automobile driver would be disadvantaged by congestion pricing. Sufficient revenues would be earned from congestion pricing, however, to ensure that middle- and lower-income groups could benefit, if the revenues were distributed with this goal in mind, for example, by using some of the revenues earned to replace other regressive taxes collected to support transportation. (Alternative uses of the revenues are discussed further below.)

Some motorists would lose.

Inevitably some motorists who pay the fees would not be in a position to receive the full benefits of the revenues that were spent. Likewise, other motorists would receive the benefits but still be disadvantaged. Some of them would switch to alternate routes, destinations, or modes to avoid the toll, but could then have trips of even longer duration and with less privacy and schedule flexibility. The revenues earned from congestion pricing can be used to ensure that all income *groups* benefit, but disadvantages to some *individuals* within these groups cannot be avoided.

Congestion pricing would reduce air pollution and save energy.

Congestion pricing has side benefits that come at no extra cost. It would reduce peak-period travel and save time, and simultaneously it would reduce air pollution and fuel consumption. The air quality and energy benefits of congestion pricing would vary widely across different areas depending on the amount of congestion, level and type of the charge, the characteristics of the transportation system, and the behavioral response. The results from simulation studies illustrate the possible benefits. A $0.06/km ($0.10/mi) congestion pricing fee charged during the peak period on all major routes in the San Francisco Bay Area is estimated to reduce total trips by 2.2 percent. Such reductions in travel are estimated to reduce emissions of precursors to ozone (oxides of nitrogen and reactive organic gases) by 3 to 5 percent, and carbon dioxide by 6.5 percent (see Table 3-1). Fuel consumption would decline by 6.5 percent. In the greater Los Angeles area, a $0.09/km ($0.15/mi) fee charged during peak periods on all congested facilities is estimated to reduce total trips by 4 percent, precursors to ozone by about 8 percent, and carbon dioxide by 9 percent (see Table 3-2). Fuel consumption would decline by about 9 percent.

Charges of $3.00/day to park in congested areas are estimated for both metropolitan areas to reduce trips by about 1.5 to 1.8 percent and reduce emissions of various pollutants by about the same percentage (Tables 3-1 and 3-2). Fuel savings would range from 1.2 to 1.7 percent.

Although the percentage reductions in total trips and emissions may appear modest to some, the air quality and energy benefits of congestion pricing strategies are much larger than those of other travel demand management policies.

The estimates presented above are for pricing throughout a metropolitan area. The first congestion pricing projects likely to occur in the United States, however, would involve individual facilities. The air quality and energy conservation benefits of such smaller-scale projects would, by their nature, be considerably smaller than those from an areawide strategy. A peak-period toll on the San Francisco–Oakland Bay Bridge, for example, is estimated to reduce regional emissions of ozone precursors by 0.10 to 0.15 percent. Even if the effect on total regional emissions is modest, pricing on individual facilities could have pronounced health benefits by reducing concentrations of carbon monoxide. When individual facilities are priced, it is very important to design projects such that they do not divert congestion to other, unpriced facilities. Doing so could reduce or negate the air quality and energy benefits of pricing on individual facilities.

The political feasibility of congestion pricing is uncertain.
Congestion pricing demonstration projects are moving forward in the United States. The Intermodal Surface Transportation Efficiency Act of

1991 (ISTEA) authorizes up to $150 million to support up to five congestion pricing pilot projects. The Federal Highway Administration has selected a proposal from the San Francisco Bay Area. Two separate, privately funded proposals are being developed in southern California.

Whether congestion pricing will prove politically feasible on a larger scale remains to be seen. Public and political perceptions about fairness and motorist resistance to paying for services formerly viewed as free continue to be significant obstacles. A complicated program for distributing revenues could be required to compensate disadvantaged groups. The substantial revenues that congestion pricing can generate provide an opportunity to improve the efficiency of the transportation system, ameliorate the negative impacts on disadvantaged groups, and result in a net benefit for society. Some individuals would still be hurt, however, and whether they would be more motivated to resist congestion pricing than the majority who would benefit will only be demonstrated in actual practice.

The adoption of congestion pricing as a demand management policy would be a drastic change from the current operation of the road system. The public would also need to be convinced that the government could successfully and fairly manage a change of this magnitude. The distrust of government prevalent in most areas of the country is an additional barrier. For this reason, there are advantages to using private-sector initiatives, such as the ones under way in Southern California, to introduce congestion pricing.

Evaluation of early projects is crucial.

If implemented, congestion pricing projects will remain controversial. The quality of the debates about these efforts would be substantially enhanced by reliable information about how traffic flows change, by careful analyses of winners and losers, and by survey research regarding motorist perceptions before and after the change. The opportunity to improve on early efforts would be greatly enhanced by such careful evaluation.

An incremental approach is appropriate.

The risks associated with congestion pricing and the nature of policy development in a pluralistic society imply that this policy will only progress in small steps. Given that congestion pricing represents a substantial change from the current operation of the road system, such small steps are appropriate. If individual projects succeed, they will help convince policy makers and the public of the benefits of congestion pricing. This process will take time, however; thus it may be many more years before congestion pricing would be applied throughout a metropolitan area in this country. Whether congestion pricing will evolve to this level will depend on how it is implemented, how well it works, and how much motorists and voters come to accept it. Only time, experimentation, and careful evaluation will tell.

RECOMMENDATIONS

Congestion pricing would improve the efficiency of the transportation system and, if the projects are properly implemented, the benefits would outweigh the costs. The benefits are sufficiently promising that local and state governments, toll authorities, and private investors should experiment with congestion pricing. Removal of federal restrictions on experimentation would create more opportunities at the local level.

Federal Government

Through the congestion pricing pilot projects authorized under ISTEA, the federal government has demonstrated considerable leadership. It is not too early, however, to consider improvements in the pilot program that would enhance the opportunity for trials of congestion pricing to go forward.

Congress should extend the pilot program when ISTEA is reauthorized in 1997.

Congestion pricing proposals have a long development period. The Bay Area project, for example, had a head start of several years in proposal development. A metropolitan area beginning to consider congestion pricing seriously in 1994 may not have a proposal together by the end of the ISTEA authorization period.

Public officials (state, local, or regional depending on who has authority) should be given discretion regarding the use of revenues collected by congestion pricing pilot projects.

ISTEA restricts the revenues earned from congestion pricing pilot projects to transportation purposes permitted under Title 23 of the U.S.

Code, essentially those eligible for federal transportation aid. This is unduly restrictive. Whereas it makes sense to condition the funding provided by ISTEA to purposes eligible under Title 23, the revenues actually earned in the pilot projects would be local funds raised from local users. A careful economic analysis could show that the most efficient and fair uses of the funds may be for purposes in addition to those allowed under Title 23. Moreover, state or local officials may well require considerable latitude in the use of the funds raised from congestion pricing in order to make a proposal politically feasible. (The committee's suggestions for appropriate uses of the funds are discussed in the next section.)

Congress should allow congestion pricing on urban Interstates or other federal-aid routes if called for in state implementation plans [as required by the 1990 Clean Air Act Amendments (CAAA)] or if local or state authorities can demonstrate the need to manage congestion on these facilities through pricing.

The CAAA requires metropolitan areas that are unable to meet CAAA standards to use travel demand management policies, such as pricing, to reduce air pollution from automobiles. Federal law, however, also significantly restricts the ability of state and local governments to impose congestion tolls on federal-aid routes, especially the Interstates. Given that urban Interstates are often the most congested facilities during peak periods, they could be prime candidates for congestion pricing. Only through the pilot program is congestion pricing now allowed on an Interstate (ISTEA authorizes experiments on no more than three Interstate routes). The current allowance for congestion pricing on Interstates could be expanded as part of the recommended extension of the congestion pricing pilot program provisions of ISTEA when this legislation is reauthorized.

The federal government should provide additional incentives to encourage pricing on more than just single facilities by giving substantial grants or additional housing, transit, or community development funds to any metropolitan area with significant congestion that is willing to experiment with broader pricing strategies, for example, a regional parking management program.

Parking pricing for congestion relief is not considered eligible as a pilot project because it can be implemented now without ISTEA. There are no federal prohibitions against the development of a regional parking management program, but implementing one would be very difficult.

For one thing, all jurisdictions within a region would have to support such an effort to avoid any one entity's gaining a competitive advantage over the others. Moreover, the development of a regional program would require considerable time to educate regional officials about the time saving and economic benefits of such a strategy, build a coalition of public supporters, and engage in the necessary public education throughout a region. Grant amounts should recognize the time and cost required up front to develop support for regional approaches to congestion pricing.

In cases where high-occupancy-vehicle (HOV) lanes on federal-aid facilities have clearly failed to induce ridesharing, or in cases where local officials can show that adding congestion pricing on existing HOV lanes would not undermine the region's HOV strategy, experiments should be allowed that would convert underused HOV lanes to tolled lanes while allowing HOV users to continue to travel at no charge.

In the first round of solicitations for congestion pricing pilot projects, several proposals requested permission to price underused HOV lanes. This option would allow solo drivers to pay a fee to use available capacity on HOV lanes and continue to allow HOVs to travel free of charge. The federal interagency review panel responsible for selecting potential congestion pricing projects rejected these proposals because the panel believed that Congress intended a narrower definition of congestion pricing. Other arguments have been advanced against this option. It may encourage solo driving, which is at variance with the goal of the CAAA to increase vehicle occupancy. It could weaken support for HOV lanes before sufficient time had passed for them to become accepted. (Where HOV lanes have been successful, it has taken several years to build regional support to allow the HOV policy to go forward and for the public to accept and begin to use the lanes.)

There would be merit, however, in allowing experiments with converting underused HOV lanes to priced lanes while letting HOV users operate free of charge in ways that would test whether congestion pricing can be a complementary policy to encourage HOVs. In these experiments, the price for solo drivers to use the HOV lane would have to be set sufficiently high that use by solo drivers did not reduce free-flow conditions for HOVs. Continuing to allow HOVs to travel free as a premium service while allowing solo drivers to pay to use the facility could better utilize existing capacity. It would thereby reduce the criticism that HOV lanes are underused and sustain support for the HOV policy. Allowing solo users to use HOV lanes in exchange for paying a fee would also obviate the complaint that something was being taken away from motorists and would thus help build support for congestion pricing. Such experiments could be restricted to cases in which it is clear that an HOV lane is underused and in which there is little prospect of inducing ridesharing or in cases in which local officials can demonstrate that the experiment would not undermine a long-term regional strategy to increase ridesharing. Allowance of such experiments may require legislation to amend ISTEA, which sets forth minimum passenger requirements for federally funded HOV lanes.

Because of the unique opportunities offered by congestion pricing projects to learn about behavioral responses to variable pricing and how they affect travel demand, the federal government should bear the bulk of the cost of extensive evaluations.

The ISTEA pilot program incorporates sufficient funds and requirements for evaluation. Other congestion pricing projects may well go forward, for example, those on the two private toll roads in Southern California; these projects deserve full-fledged evaluation as well. Extensive, multiyear evaluations are expensive. They could easily cost several million dollars, but they are also well-justified public expenses. Metropolitan areas throughout the nation could learn the efficacy of such projects through evaluation research. Because of the broad applicability of evaluation research, the expense should be borne by the federal government rather than states or localities.

Matching project development grants should be made available to local governments, states, toll authorities, and metropolitan planning organizations funded out of the congestion pricing pilot program section of ISTEA.

The federal government should give grants to local governments, states, or MPOs that would encourage these entities to examine the potential of congestion pricing, develop detailed proposals, and begin testing the practical and political feasibility of congestion pricing. In order to ensure that there is genuine local interest, the grantees should be willing to invest their own funds at a level that would equal that of the federal government.

Federal law should treat the tax-exempt status of parking and transit subsidies equally and should require employers who provide parking subsidies to give employees the option of taking this subsidy in the form of cash.

Current federal tax law has a built-in bias that encourages employers to provide parking as a tax-exempt benefit and thereby encourages driving to work. The current tax-exempt status of employer-provided parking is limited to $155/month, whereas transit and vanpool subsidies are limited to $60/month. No tax benefits are provided to employers for employer-provided subsidies to encourage employee ridesharing, bicycling, or walking to work.

California has enacted an innovative law that requires employers who lease parking from third parties to give employees the option of taking the parking benefit as a cash option (referred to as "cashing out" employee parking). This law has been designed so that there are few, if any, disadvantages and many advantages. Employees who want to continue to drive and park are allowed to do so. Employees who can find other options for getting to work can take the benefit in cash. Employers bear little or no additional expense, and governments receive taxes on the benefit when given in cash rather than in kind. This policy is another incremental step toward reducing the incentives to drive during peak periods. The federal government could encourage the adoption of the cash-out option by defining qualified parking as a tax-exempt fringe benefit. Such parking would be defined as that parking provided to an employee on or near the business premises of the employer if the employer offers the employee the option to receive, in lieu of the parking, the fair market value of the parking, either as a taxable cash commute allowance or as a mass transit or ridesharing subsidy. The Clinton administration endorsed this proposal in its Climate Change Action Plan of October 1993.

State and Local Governments

The substantial revenues that congestion pricing can generate could be used to replace other taxes, compensate those who lose, or operate the transportation system more effectively. The most effective and efficient uses of the funds, however, will depend on local circumstances. The committee does not have a specific recommendation to offer on the use of revenues; authorities that develop congestion pricing proposals should consider the following suggestions.

To ensure that all groups benefit and to help clarify that congestion pricing is not just another tax, some of the congestion pricing revenues could be returned to road users by reducing taxes that support transportation (gasoline, property, and sales taxes in some areas). This would offset criticism of congestion pricing, and the replacement of regressive taxes used to support transportation would help address equity concerns, particularly in areas where transportation is supported from sales taxes. This revenue-neutral approach is desirable but may not always be achievable. In some metropolitan areas, for example, such an approach may not benefit users most directly affected. This would be the case particularly when congestion pricing is limited to a single facility.

In the case of pricing on a single facility, the revenues could be used to improve alternative facilities or to subsidize transit services in the same corridor. Once implemented, congestion pricing might well improve bus transit service by reducing congestion. This would allow bus service to become faster and could allow for more frequent or expanded service. Furthermore, the imposition of a fee on motorists would induce some to shift to transit, which would thereby improve transit revenues and would make transit more cost-effective by using available capacity more efficiently. In order to provide for improved service when congestion fees are imposed, there would be justification for subsidies to purchase additional buses. The availability of alternate routes or modes for those unwilling to pay a congestion fee on a priced segment will be critical to the public acceptance of projects; thus authorities may need to provide for such alternatives in advance, with the cost reimbursed from future revenues. Subsidies for transit services should be targeted at services that would directly benefit residents in negatively affected geographic areas or groups who would use this mode. Revenues raised by congestion pricing in general should be returned to the geographic area from which they are raised.

State governments should adopt statutes similar to the California law that requires "cashing out" employee parking.

As noted earlier, the California law applies only to employers who lease parking from third parties separately from office space. State legislatures could take the cash-out option one step further by requiring leases that combine building space and parking to be separated when they are renewed; this would then make a higher proportion of leases subject to the cash-out option.

RESEARCH PRIORITIES

Careful and extensive evaluation of any congestion pricing program introduced in the United States is the highest priority for research. Research is also encouraged in other areas:

- The impact of congestion pricing on business logistics and commercial carriers;
- The extent to which transit services and revenues could be improved as a result of congestion pricing and how this might benefit lower-income users;
- Development of improved models for simulating household travel changes in response to pricing and other travel demand management strategies;
- Improved measures of congestion;
- Efficiency and productivity benefits of congestion pricing;
- Development of a program to ensure that the United States learns from current and emerging experiments with road pricing in other parts of the world;
- Measurement of long-term land use changes that might occur in response to congestion pricing;
- Studies of how the benefits and burdens of policies such as congestion pricing shift over time through labor, land, and retail markets;
- Constituency building and the local politics of implementation; and
- The efficacy of distributing tradable permits to all motorists for driving during peak periods as an alternative to charging congestion tolls (this option would allow motorists to be "bought off" of congested routes rather than being "tolled off").

1

Introduction

Even while concentrating on the traffic, drivers on routine automobile trips to work listen to the radio, think about the day ahead, or perhaps just daydream. The traffic moves at its normal pace. Drivers accustomed to the same route make their usual turns or lane changes. Jangling radio commercials interweave with the Top Forty. But the lane ahead is suddenly a long string of glowing red taillights. Hearts sink. Congestion. Blood pressures rise. Delay. Traffic slows to a standstill. . . . Perhaps this time it's a minor problem that will clear up shortly, but after five minutes of creeping along, hope gives way to anxiety about being late. After ten minutes, anxiety gives way to anger about the wasted time and work schedules disrupted.

This story sounds familiar to us all. Millions of drivers endure frustrating delays every day. Viewed from a helicopter hovering above the Bay Bridge in Oakland and as far as the eye can see, cars creep along 12 lanes of incoming traffic during the morning peak period. The traffic queues south of Washington, D.C., where Interstate 95 meets the Capital Beltway can stretch for 35 km (22 mi) during peak-hour traffic jams. Even before the 1994 earthquake, the "rush hour" on the Santa Monica Freeway lasted all day. The peak morning and evening traffic flow on the Long Island Expressway can be measured in inches per minute.

Congestion not only is frustrating, it is costly. The biggest single expense of congestion is motorist time lost because of delay, but stop-and-go traffic also wastes fuel. The cost of delay in U.S. urban areas, based on estimates of motorist value of time and wasted fuel, totaled $43 billion in 1990 (Shrank et al. 1993). The cost would rise even more if it included the increased air pollution from the emissions of idling and accelerating engines that accompany stop-and-go traffic. These figures also do not measure how congestion affects the cost of freight movements and how this, in turn, makes it more difficult for U.S. products to compete in world markets.

As described later in more detail, traffic congestion in and around major metropolitan areas has persisted during the past two decades despite the expenditure of billions of dollars on urban freeways and public transportation systems. Nor have various forms of recent traffic demand management techniques offered substantial relief. The lack of success with both traditional and new methods, combined with fiscal restraints and environmental restrictions on highway expansion, has renewed interest in an old idea—congestion pricing.

Congestion pricing could significantly reduce traffic congestion. Congestion pricing has a long history in the academic and professional literature and has been applied with success in Singapore since 1975. As discussed in Chapter 2 and Appendix B, it was possible to introduce congestion pricing swiftly and without controversy in Singapore in large part because the island state is not a Western-style democracy and is without the jurisdictional complexities characteristic of U.S. metropolitan regions. The pricing scheme itself is fairly simple. Drivers who wish to drive during the peak period are charged for the delays they impose on each other. When faced with a congestion fee, some drivers will decide that it is not worth it to drive (or drive alone) during the peak period. They may choose to drive at a different time, share a ride in a carpool, use transit, or make some other arrangements. Drivers willing to pay the fee will save time and the reduced congestion will reduce automobile emissions. Although congestion pricing increases out-of-pocket expenses and causes some commuters to change their behavior, society as a whole will benefit because of the resulting time savings.

Congestion pricing, however, faces a considerable political challenge. Motorists are accustomed to traveling on roads without paying tolls. Many are concerned that congestion fees would be unfair to low-income drivers or to drivers who would not like to pay the fee but would have no other choice. When raised in the United States in the past, congestion pricing proposals were summarily dismissed.

Despite its past lack of acceptance, congestion pricing is subject to renewed interest. Environmentalists increasingly see congestion pricing as a potential tool for meeting state and federal air quality standards, particularly the more stringent standards in place in California (Cameron 1991). Further impetus for congestion pricing—indeed for this study—is provided by the Intermodal Surface Transportation Efficiency Act of 1991

(ISTEA), which allows up to $25 million per year through 1997 to help fund up to five congestion pricing pilot projects. One congestion pricing pilot project was approved in 1993.

Congestion pricing is also being reconsidered because of growing awareness of revenue shortfalls for capital investment from traditional transportation taxes and increased acceptance of toll financing (Small et al. 1989). For example, because of a lack of funds and an inability to respond to the demand for new facilities, the state of California is allowing seven new toll roads to be built, two of which are planning to rely on peak-period fees.

IMPETUS FOR CONGESTION PRICING

Congestion

Motorists do not like congestion and believe that it is getting worse. A recent nationwide poll found that nearly two-thirds of central-city and metropolitan-area residents believe that congestion is a major problem (Apogee Research, Inc. 1990). Half of the respondents to this survey reported that congestion had become "a lot worse" compared with the situation five years ago. Data available on the use of specific facilities support this view. Although the capacity of major arterial highways increased 25 percent between 1980 and 1991, the volume of travel during that same period increased almost 60 percent. As a result, the percentage of travel on urban Interstates operating in or near congested conditions has increased considerably (Figure 1-1). (Other available trend data on congestion are reviewed in Appendix A.)

Efficiency

Economists have long advocated congestion pricing because it would use the existing transportation system more efficiently (Walters 1961; Vickrey 1959; Mohring 1965; Keeler and Small 1977; Hau 1992). Most metropolitan areas would have sufficient highway capacity if the daily use of the system could be averaged over the hours of the waking day. Highway use, however, like use of utilities, restaurants, seasonal resorts, and airlines, is characterized by considerable differences between peak and off-peak demand. Highways in most areas are typically congested during the morning and evening peak hours and underused the rest of the day. (Many urban

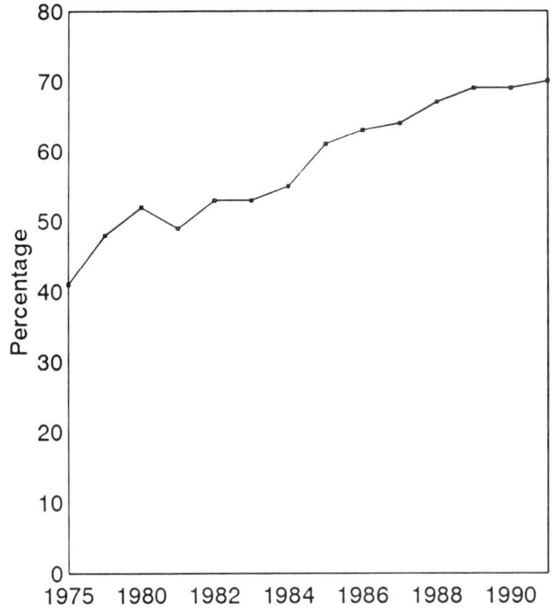

FIGURE 1-1 Percentage of peak-hour urban Interstate travel occurring in congested conditions (DOT 1992). (As defined in this figure, congestion occurs when traffic meets or exceeds a volume-to-service-flow ratio of 0.80. See Appendix A.)

areas, however, are now also confronted by a midday peak, and some routes within metropolitan areas are congested throughout the day.) By shifting some of the demand away from peak periods, the traffic would flow much more smoothly during peak periods and the facility would be more efficiently used in the off peak. As well as easing congestion on an existing facility, congestion pricing would also maximize the efficiency of use.

Although economists' espousal of congestion pricing has gone unheeded, one of their predictions has been borne out. When scarce and valued goods, like highway capacity, are given away, queues of people who want them will form. The nation has invested billions of dollars in the existing highway and transit system, often justified on the basis of reducing congestion (Altshuler 1979). Yet there is little evidence that urban traffic congestion has been eased.

New construction alone will not significantly ease congestion in high-growth parts of the country because of changes in traffic behavior that occur when new capacity is added (Downs 1962). When motorists perceive that congestion has lessened during the peak, they change the timing of trips or make trips they formerly did not make to take advantage of the new capacity. Some former carpoolers will revert to driving alone. This phenomenon is referred to as latent demand. In addition, motorists who formerly took alternate routes during the same time period will soon discover that the new capacity offers a quicker route, and they will shift to the new route. Before long, the latent demand congests the new capacity during the peak. Over the longer term, the provision of new capacity also stimulates additional residential and commercial development that brings new motorists onto the road. Although the new road will serve more vehicles than before and will save time for many commuters compared with their trip times before the new capacity, when the new capacity becomes congested, motorists become frustrated with the delays and the stop-and-go traffic increases air pollution. Congestion pricing is proposed as the only demand management technique—short of administrative rationing—that can keep latent demand from congesting new facilities (or recongesting other facilities after some other measure has helped alleviate the peak demand).

Restrictions on Expanding Highway Capacity

During the last four decades, the conventional response to congestion has been to increase capacity. However, both ISTEA and the Clean Air Act Amendments (CAAA) of 1990 contain strict prohibitions on the addition of new highway capacity in areas failing to meet federal air quality standards. As of October 1993, 41 areas were designated as being in nonattainment for carbon monoxide (CO), for which motor vehicles are responsible for 80 percent of emissions, and 94 areas were designated as being in nonattainment for ground-level ozone (EPA 1993).

Instead of providing new capacity, those areas not in compliance with the CAAA must propose specific measures for reducing automobile travel through measures such as trip reduction ordinances, employer-based transportation management, transit improvements, pricing, traffic flow improvements, and parking management. Employers with more than 100 employees in the 10 metropolitan areas rated as in "severe" or "extreme" nonattainment are required to submit plans by 1994 that will result in reductions in the number of employees driving to work alone. An employee trip reduction program required under California law is already in effect in Southern California. Early experience with California's employee trip reduction program indicates that such programs are unpopular with companies and at best result in modest traffic reductions in the region (Giuliano and Wachs 1992). The prospect that employee trip reduction programs and other traffic restraints would be imposed in the San Francisco Bay Area led business leaders to urge consideration of congestion pricing as a more palatable alternative (Bay Area Economic Forum 1990).

Advances in Technology

Congestion pricing was dismissed by some decision makers in the past partly because having motorists line up at toll booths to pay tolls would itself create congestion. This problem has given way to advances in automated toll collection. Some toll agencies in the United States are already collecting tolls electronically at considerable cost savings compared with the cost of manual and automatic toll booths. Computers at toll collection sites communicate with electronic tags in vehicles without requiring motorists to stop (Figure 1-2). These tags, some of which are the size of a credit card and about twice as thick, are placed on the dashboard or windshield. Clients can prepay for a period and have their accounts automatically debited with each use or can be billed for use on a monthly basis.

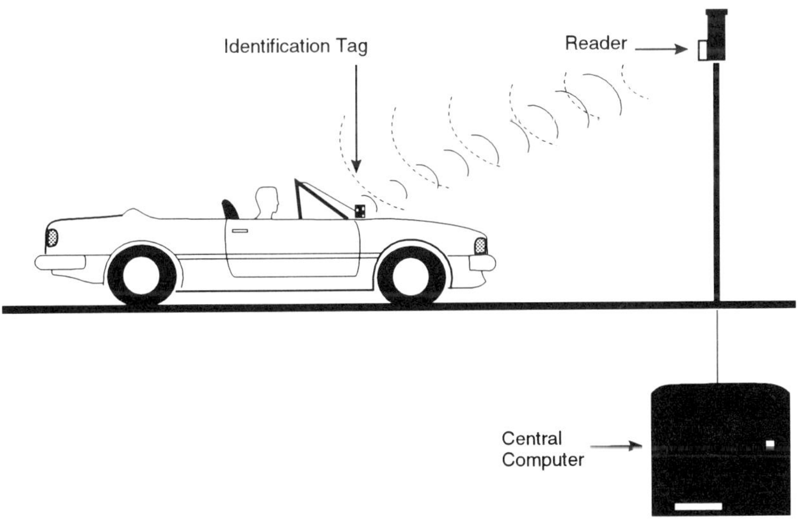

FIGURE 1-2 Electronic toll collection.

Toll agencies have set up procedures to protect the privacy of users; further advances in technology will enhance privacy even more (Pietrzyk, Vol. 2).

Making the Automobile User Pay

Congestion pricing, together with various fees or taxes on automobile use to reduce vehicle emissions, is being advocated by both business and environmental groups as a means of curbing growing automobile demand, reducing air pollution, and better balancing the benefits of automobile use with its social and environmental costs (Bay Area Economic Forum 1990; Cameron 1991). Most road users believe that they have paid for their road use through gasoline taxes. User taxes, however, cover only about 80 percent of total, direct roadway costs at the federal level and even less at the state and local levels (Small et al. 1989; DOT 1992). Direct roadway costs are defined to include maintenance of roads and roadway structures, construction, traffic police, street lighting, and other highway-related costs. The share of direct costs at the local level is much lower; in Wisconsin cities, for example, about half of the total direct cost of highway transportation is paid from local property taxes (Hanson 1990).

Most automobile users also receive a parking subsidy in the form of employer-provided parking, which is a fringe benefit that is not subject to federal or state income tax (Shoup, Vol. 2). According to the Nationwide Personal Transportation Survey, almost all employees receive free parking, which encourages solo driving (Shoup, Vol. 2).

Highway users also impose indirect costs on society, including air pollution, noise, traffic accidents, and the delays that users impose on one another during peak periods. Efforts to estimate the total indirect costs that highway users impose on society are usually controversial because of the difficulty of valuing environmental damage, pain and suffering, and other nonmonetary losses. Despite the controversy, few would dispute that these costs are substantial.

Through strategies that could include tolls on roads or bridges, fees to enter congested areas, or changes in the structure of parking and transit pricing, congestion pricing would charge a premium to motorists who wish to drive on normally congested routes during peak travel periods. The subsidies to peak-period commuters have been characterized by the economist John Meyer as follows:

> Commuters going to work in major central business districts in the United States in their own motor vehicles directly pay for only about 25

> percent of the total cost of their transport. The other 75 percent is typically borne by their employers (e.g., providing free parking), by others (in increased congestion, reduced safety, etc.), by fellow workers or residents (in air or noise pollution, etc.) and by governments. (Johnson 1993, 10–11)

Supporters of congestion pricing and fees on automobile use believe that if highway users paid directly for more of the costs they impose, the demand for automobile use, or the negative impacts of automobile travel, would be reduced.

RESISTANCE TO CONGESTION PRICING

Road pricing, of which congestion pricing is a primary ingredient, has been a matter of policy debate in the United States before. During the expansionist period of highway construction following passage of the Federal Highway Revenue Act of 1956, it was discussed as an alternative approach for defining long-run investment needs and for solving the urban congestion problem, but was never acted upon. Following the passage of the Clean Air Act (CAA) of 1970 and the energy crisis of 1973, federal transportation policy shifted somewhat away from an emphasis on highway construction toward concern for achieving greater efficiency in urban transportation (Altshuler 1979). Federal funding for public transit increased and the transportation system management (TSM) concepts for increasing the efficiency of highway use came into vogue. Interest was further stimulated by the Environmental Protection Agency (EPA) transportation control plans for those urban areas that failed to meet the ambient air quality standards of the CAA. Congestion pricing was one of the major policy instruments considered during this period. For example, beginning in 1976, then-Secretary of Transportation William Coleman offered several cities federal funding if they would participate in studies of congestion pricing.

Although subject to serious intellectual discussion, actual congestion pricing proposals were greeted by policy makers with skepticism if not downright hostility. The Bureau of Public Roads (forerunner of the Federal Highway Administration) rejected road pricing proposals as unworkable (St. Clair 1964). Congress prohibited the EPA from using areawide parking pricing, one approach to congestion pricing, when EPA moved toward this proposal in 1973. Although a few cities gave Secretary Coleman's invitation to study congestion pricing serious consideration, no local government participated (Altshuler 1990; Higgins 1979, 1986). The mayors and council members who turned down Coleman's offer cited concern about the effect on low-income individuals, the reaction of users who had become accustomed to "free" highways, and adverse consequences for struggling downtown businesses.

The reasons for rejection of congestion pricing in the past have not changed. Any shift from the current system of financing and using the transportation system toward more marketlike mechanisms can be expected to engender public and political resistance (Wachs 1992). Concern that the poor would be less able than the middle class or the rich to pay congestion fees is often held up as an important political barrier to the acceptance of congestion pricing. (The unwillingness of some in the middle class to pay higher out-of-pocket expenses may be the more salient political barrier.) New road use charges of any sort are widely viewed as new taxes by a citizenry that holds the government in low esteem. In addition, some commercial interests in congested areas that might be subject to congestion pricing may be more worried about the possible short-term costs than the possible long-term benefits.

The political difficulties of congestion pricing have not gone away, but the forces in favor of congestion pricing have increased. Whereas local political leaders eschewed congestion pricing in the past, support is growing in some areas, particularly in California. Although congestion pricing proposals are proceeding in a small-scale, incremental fashion, that they are proceeding at all is a substantial change from this policy's fate when last seriously considered in the United States.

OUTLINE OF REPORT

Volume 1 is an overview of the papers contained in Volume 2 and a summary of the committee's view of the potential for congestion pricing in the United States. Chapter 2 provides a brief overview of the theory and definition of congestion pricing, experience abroad, and current proposals for the United States. Chapter 3 summarizes the best available estimates of how congestion pricing might affect travel, different groups of motorists, air quality, energy, and urban form. Chapter 4 provides an overview of the administrative and political barriers to congestion pricing and how they might be overcome. The design of congestion pricing pilot projects and importance of evaluating any proposals that might be implemented are discussed in Chapter 5. The committee's conclusions are contained in Chapter 6.

The appendixes contain more detailed material about some of the same issues covered in the report. This material is provided for readers interested in delving more deeply into available measures of traffic congestion (Appendix A) and the theory of congestion pricing and studies of its possible effects in the United States (Appendix B).

Volume 2 contains the papers commissioned by the committee, most of which were presented and discussed at a symposium held in June 1993 at the National Academy of Sciences. Before publication, the authors revised their papers on the basis of the discussion. Taken together, the papers provide a rich array of information about individual case studies from around the nation and thoughtful analyses by individual scholars about many of the critical issues surrounding congestion pricing.

REFERENCES

ABBREVIATIONS

DOT	U.S. Department of Transportation
EPA	Environmental Protection Agency

Apogee Research, Inc. 1990. *The Road Information Program (TRIP) National Transportation Survey: 1990 Poll Results.* Summary Report. Road Information Program, Washington, D.C.

Altshuler, A. 1990. Discussion of C. Winston, How Efficient Is Current Infrastructure Spending and Pricing? In *Is There a Shortfall in Public Capital Investment?* Proceedings of a Conference Sponsored by the Federal Reserve Bank of Boston, Conference Series No. 34, pp. 206–213.

Altshuler, A. 1979. Congestion. In *The Urban Transportation System: Politics and Policy Innovation*, Chap. 9, MIT Press, Cambridge, Mass.

Bay Area Economic Forum. 1990. *Market Based Solutions to the Transportation Crisis: Incentives to Clear the Air and Ease Congestion.* San Francisco, Calif.

Cameron, M. 1991. *Transportation Efficiency: Tackling Southern California's Air Pollution and Congestion.* Environmental Defense Fund, New York City; Regional Institute of Southern California, March.

DOT. 1992. *Highway Statistics, 1991.* Federal Highway Administration.

Downs, A. 1962. The Law of Peak-Hour Expressway Congestion. *Traffic Quarterly*, Vol. 16, July, pp. 393–409.

EPA. 1993. *National Air Quality Emissions Trends Report, 1992.* Report EPA 454/R-93-031. Office of Air Quality Planning and Standards, Research Triangle Park, N.C.

Giuliano, G., and M. Wachs. 1992. A Comparative Analysis of Regulatory and Market-Based Transportation Demand Management Strategies. In *Papers*

Presented at the Congestion Pricing Symposium, June 10–12, FHWA and FTA, U.S. Department of Transportation, pp. 6-1 to 6-15.

Hanson, M. 1990. Automobile Subsidies and Land Use: Estimates and Policy Responses. *Journal of the American Planning Association*, Vol. 58, No. 1, pp. 60–71.

Hau, T. 1992. *An Economic Analysis of Road Pricing: A Diagrammatic Approach.* Policy Research Working Papers. WPS 1070. The World Bank, Washington, D.C.

Higgins, T. 1979. Road Pricing—Should and Might It Happen? *Transportation*, Vol. 8, pp. 99–113.

Higgins, T. 1986. Road Pricing Attempts in the United States. *Transportation Research*, Vol. 20A, No. 2, pp. 145–150.

Johnson, E. 1993. *Avoiding the Collision of Cities and Cars: Urban Transportation Policy for the Twenty-First Century*. The American Academy of Arts and Sciences, Cambridge, Mass.

Keeler, T., and K. Small. 1977. Optimal Peak-Load Pricing, Investment, and Service Levels on Urban Expressways. *Journal of Political Economy*, Vol. 85, No. 1, pp. 1–25.

Mohring, H. 1965. Urban Highway Investments. In *Measuring Benefits of Government Investments* (R. Dorfman, ed.), The Brookings Institution, Washington, D.C., pp. 231–275.

Shrank, D., S. Turner, and T. Lomax. 1993. *Estimates of Urban Roadway Congestion, 1990.* Report 1131-5. Texas Transportation Institute, College Station.

Small, K., C. Winston, and C. Evans. 1989. *Road Work: A New Highway Pricing and Investment Policy*. The Brookings Institution, Washington, D.C.

St. Clair, G. 1964. Congestion Tolls—An Engineer's Viewpoint. In *Highway Research Record 47*, HRB, National Research Council, Washington, D.C.

TRB. 1985. *Special Report 209: Highway Capacity Manual.* National Research Council, Washington, D.C.

Vickrey, W. 1959. Statement on the Pricing of Urban Street Use. Joint Committee on Washington Metropolitan Problems, U.S. Congress. *Hearings on the Transportation Plan for the National Capital Region*. Nov.

Wachs, M. 1992. Introductory Remarks. Presented at Symposium on the Role of Pricing and Market-Based Strategies, Lake Arrowhead, Calif., Sept. 7.

Walters, A. 1961. The Theory and Measurement of Private and Social Cost of Highway Congestion. *Econometrica*, Vol. 29, No. 4, pp. 676–699.

2
Theory and Experience

Peak-period pricing of consumer goods and services is common throughout the economy. Resorts cost more during the high season. Flights scheduled at times preferred by business travelers (early morning and late afternoon on weekdays) cost a premium. Long-distance telephone calls cost more if made during working hours than if made late in the evening. As discussed in the first section of this chapter, there is a strong rationale for peak-period pricing of scarce goods subject to peak demands—doing so ends up costing society less of its available resources than either rationing use or expanding capacity to meet peak demand. In the subsequent section, various approaches are described that can be used to apply congestion pricing on road systems. The final section is a review of experience to date with congestion pricing worldwide.

THEORY

Whenever a scarce and valued good is free or underpriced, demand will outstrip supply. This is illustrated for the case of road use by the queues and traffic jams that occur when the number of motorists attempting to use a section of roadway at the same time exceeds the road's capacity. Long-distance telephone service is another example of a service that experiences peaking. Expanding capacity to meet peak demand results in wasteful excess capacity unless the peak users are charged the full cost of the expansion. Long-distance telephone service allocates the demand with a market mechanism by charging a premium for a call during the peak and by offering discounts during the off peak. By shifting a small portion of the peak demand into the off peak, excess capital investments are avoided and society receives the service it requires at a lower cost. Consumers appear to accept a market-based system for allocating demand in long-distance telephone service with some equanimity, and experience with peak-period pricing in air travel suggests that these policies work and that they are accepted by the populace (Gillen, Vol. 2).

In the case of roads, however, demand is allocated by congestion. Everyone who wants to travel in the peak tries to do so, and the excess demand causes congestion. Economists and travel demand forecasters believe that travel decisions are governed by the "full price of travel," which includes the out-of-pocket expense plus the value of the time it requires. Given that even so-called free road use comes with significant congestion during the peak, it appears reasonable to ask why motorists willingly endure the delays that occur when they travel at the most congested times. Indeed, the available evidence suggests that when congestion gets bad enough, some travelers shift the timing of their trips. These shifts, however, are not enough to alleviate the crush. The reason that shifts in behavior by drivers confronted with congestion will never be sufficient to reduce congestion (without some form of pricing or rationing) is that motorists using the highways during the peak do not have to pay for the delays they cause each other.

To illustrate this problem, imagine an 8-km (5-mi) long, high-speed segment of highway with a flow rate of 1,999 vehicles per hour in each lane. Such a flow rate would be quite high and nearing the point when the design capacity of the facility would be exceeded. As demand approaches capacity, the throughput of the facility begins dropping disproportionately, which causes delay to mount rapidly. Standard assumptions about vehicle speed under these conditions imply that the trip for the average vehicle will require 25 min (Wohl and Hendrickson 1984, 290). The next vehicle to enter the traffic stream will have a slightly longer trip than 25 min but would add about half an hour of delay to be shared by all the following motorists. This aggregate half-hour of added delay turns out to be fairly small when spread over the thousands of following motorists in the following hours (just 1 second each in this example), but these increases mount up as additional drivers enter the traffic stream. If 400 motorists were added to this traffic stream, the first few additional motorists would be able to complete their trips by slightly more than the average amount of time, but by the time the entire 400 extra motorists had been added, the following traffic would need 34 min (an additional 9 min each) to complete their 8-km (5-mi) trip. In this example, a 20 percent increase in volume results in a 36 percent increase in the average time required to complete the trip. Congestion pricing would allow motorists to make peak-period trips under less congested conditions, but only if they are willing to pay for the delays they impose on each other.[1]

Making motorists pay for the delays caused others by their trips rather than just their personal costs would lead some to make other choices. They might decide to travel at another time, share a ride, take an alternative mode of transportation, or forgo the trip. The price would be set at a level that reduces congestion to its most efficient level, which can be shown to be the full monetary and time cost for using a segment of road capacity during the peak period (Appendix B).

Congestion pricing might also induce other changes that would further reduce the demand for peak-period travel. Employers might shift their hours of operation, allow greater use of flex time by employees, or allow more telecommuting. In contrast, however, employers could also engage in efforts that would defeat the intent of congestion pricing, for example, by subsidizing the congestion fees paid by some employees (Deakin, Vol. 2). Over the longer term, pricing strategies might also affect land use patterns. These potential secondary effects, both short- and long-term, are poorly understood. (They are discussed in more detail in Chapter 3 and in the paper by Deakin in Volume 2.)

Congestion could presumably also be controlled by regulation. Driving could be restricted by the same kinds of rationing schemes used during the fuel shortages of the 1970s. Motorists with license plates ending in even numbers could drive on even-numbered days and motorists with license plates ending in odd numbers could drive on odd-numbered days. Alternatively, trucks could be restricted from operating in congested corridors during peak hours, as has been proposed in some urban areas. Congestion pricing is advocated over a regulatory approach because allowing individ-

[1] Motorists impose different costs on one another and on society. The text refers to the delay costs, which are the extra (marginal) time costs created when demand for road capacity exceeds supply. Other road user costs include wasted fuel and extra air pollution, the latter of which is a cost borne by the public. Congestion fees, as used in this report, are defined as marginal time costs. Full road pricing could be defined to include environmental, safety, and other costs of highway transport.

uals to choose whether to pay a congestion fee will result in a more efficient outcome.

APPLICATION

Congestion pricing could take several forms. The most straightforward example is to add an extra fee to an existing toll or to add a peak fee on an untolled route or bridge. The charge itself could be a simple peak/off-peak differential, or it could vary according to the demand imposed on a facility at a specific time. In more complex forms, it could apply to traffic on all routes entering a congested area, and the charge could vary according to the congestion on specific routes and the distance traveled. Alternatively, all motorists parking in a congested area could be assessed an extra fee to encourage them to use other modes or to share rides.

In practice, congestion pricing could take six basic forms (Gomez-Ibanez and Small forthcoming):

- **Point pricing**, in which a highway user passing a point at a specific time is charged a fee for passing that point regardless of the distance traveled on a specific route (Figure 2-1);
- **Cordon pricing**, in which users wishing to enter a congested area are charged fees at each entry point into the area (Figure 2-1);
- **Zone pricing**, in which users traveling within a cordoned area also pay a fee;
- **Higher charges for parking** in congested areas, with particular emphasis on parkers traveling during the most congested period;
- **Charges for distance traveled** within a congested area or on a congested route (Figure 2-2); and
- **Congestion-specific charges**, in which users would be charged for both time spent and distance traveled.

Currently the only forms of congestion pricing in operation are cordon pricing in Singapore and charges for distance traveled on the A-1 toll road outside Paris. Cordon pricing is also being applied in three Norwegian cities, but because the tolls do not vary by time of day (with a minor exception in Trondheim), these are not considered congestion pricing.

The active proposals for congestion pricing in the United States are for point pricing. A hybrid version of point pricing has also been proposed in the United States; in this version, solo drivers would be provided with the option of paying to travel on underused high-occupancy-vehicle (HOV)

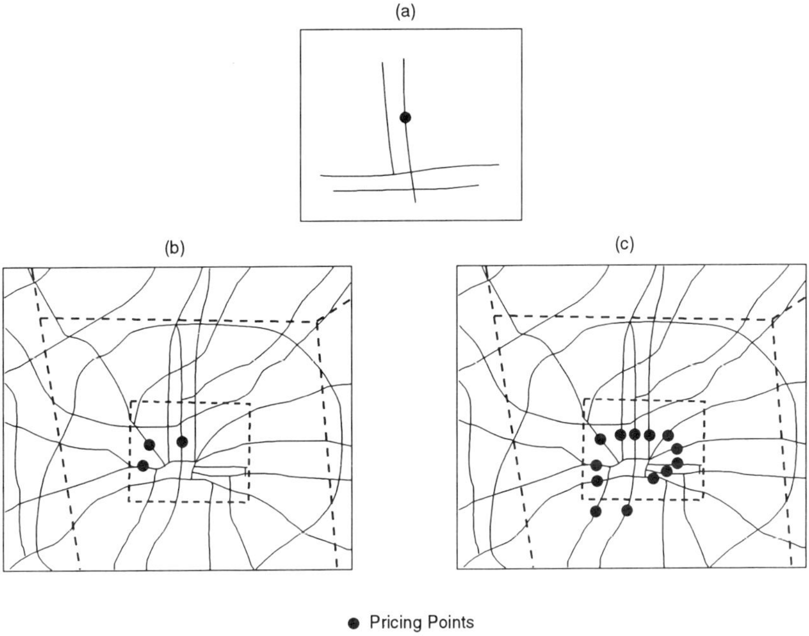

FIGURE 2-1 Examples of point pricing of traffic entering the central area. Point pricing could apply on (*a*) an individual route, (*b*) only the most congested routes, or (*c*) all the routes serving a central area. The first case would be facility pricing, the second would be an extended version of facility pricing, and the last would be a form of cordon pricing.

lanes. Cordon pricing is planned for Stockholm beginning about 1997, and point pricing is being actively considered in the Netherlands. A multiple-cordon pricing system was studied for Hong Kong and is being studied for London (Gomez-Ibanez and Small forthcoming). Rapid innovations and applications in electronic toll collection technologies are making the more complex pricing versions (e.g., zone pricing and charges for distance traveled) increasingly feasible (see papers by Pietrzyk and by May in Volume 2).

EXPERIENCE

Singapore

A simple form of congestion pricing has been in place in Singapore since 1975 (Hau 1992). Motorists wishing to enter the core area of the city covering 5 km^2 (about 2.4 mi^2) during the morning and evening peaks

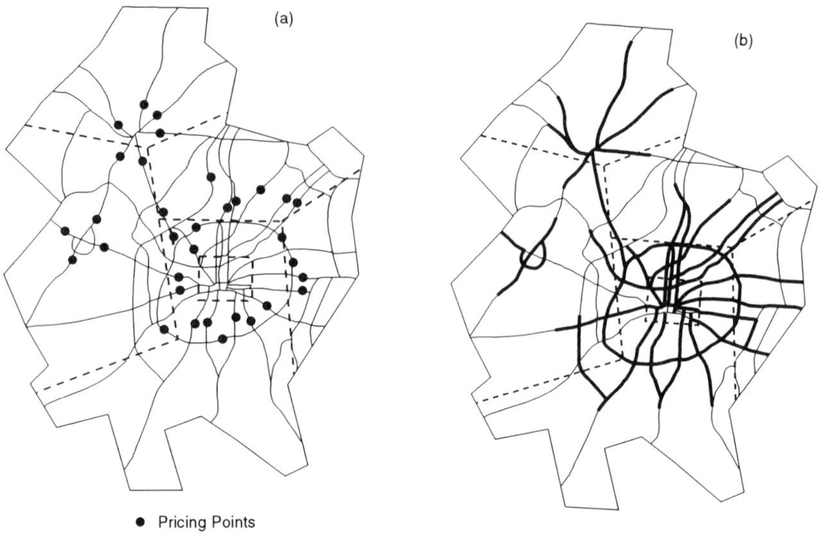

FIGURE 2-2 Examples of regional pricing. (*a*) Point pricing on all major radial routes serving the central area and major activity centers and a beltway surrounding the central city. (*b*) Charges for distance traveled on all major radial routes and a beltway.

must display a windshield sticker. The sticker currently costs S$3.00 (about U.S.$1.80) per round trip. The stickers are enforced at 33 access points into the zone; these access points are not always monitored, but noncompliance has remained below 1 percent. Shortly after this system was imposed, the number of automobiles entering the central part of the city during the restricted hours declined by 44 percent. Almost 20 years later, and after several modifications to the program, automobile trips into the center city during the morning peak period remain some 25 percent below the 1974 level (Gomez-Ibanez and Small forthcoming).[2]

The Singapore program, however, did have some important design flaws. The road that circled the city could not accommodate the traffic diverted from traveling through the city. The time losses due to congestion caused on the ring road initially negated the time savings for travelers entering the city core; this problem was later alleviated by adding lanes to the ring road (Gomez-Ibanez and Small forthcoming).

[2] As described in more detail in Appendix B, the evening peak period was not subject to a congestion fee and remained quite congested until a fee was imposed in 1989.

Despite the design flaws, the restraint on traffic growth into the urban core is all the more remarkable considering that during this same period Singapore's population growth exceeded 23 percent and employment grew by over 79 percent. Although Singapore is very different from metropolitan areas in the United States and the program design has some important faults, Singapore's experience indicates the effects that congestion pricing can have on reducing peak-period demand. No other travel demand management policy tried in the United States short of wartime gasoline rationing has been shown to have such a significant and lasting effect on automobile use. Singapore, of course, is not a Western-style democracy, and the fact that areawide congestion pricing could be introduced in Singapore is not evidence that it could be introduced and implemented in the United States.

France

Congestion pricing also has been shown to be effective on an intercity highway in France. A simple congestion pricing system has been implemented on a toll road outside Paris (the A-1) to deal with the weekend peak demand caused by the return of city residents on Sunday afternoons from weekend retreats (Gomez-Ibanez and Small forthcoming). The usual toll applies up until 2:30 in the afternoon. From 2:30 to 4:30 p.m. a discount is offered that is 25 to 50 percent (depending on the length of the trip) below the normal toll. From 4:30 to 8:30 a premium is charged that is 25 to 50 percent higher than the normal toll. After 8:30 and until 11:30 p.m., the discount is again offered, and after 11:30 the normal toll applies. These price differentials have caused enough travelers to shift their behavior to end the excessive delays. They were specifically designed to be revenue neutral; that is, the toll authority designed the premium and discount to ensure that its revenues would remain constant and that the congestion fee would simply smooth traffic flow. Little traffic has been diverted to alternate routes, perhaps because they are not built to freeway standards.

Hong Kong

Hong Kong conducted an extended experiment with electronic road pricing technology in 1983 to 1985. In order to test the accuracy and reliability of the equipment (no congestion fee was charged), over 2,500 vehicles were outfitted with transponders, which were identified and read whenever they passed over electronic toll points imbedded in the pavement. By all accounts the technology, which was less sophisticated than current systems, performed quite well. Vehicles were recorded for over 99 percent of crossings of the electronic toll points they did make and were not recorded for crossings they did not make (errors were kept below a rate of 1 in 10 million) (Dawson and Catling 1986). When the government allowed neighborhood councils to vote on whether to move to implementing congestion pricing, however, the proposal was soundly defeated. Some authors claim that the public resented the invasion of privacy, but Borins (1988) suggested that the opposition involved more than just the congestion pricing proposal itself. This was the first time the British Governor had allowed the councils to vote on any issue; the explanation of the program and its benefits was clumsily conveyed to the public, and it occurred just at the time the British had agreed to return Hong Kong to China. The neighborhood councils may well have rejected any proposal the government brought forward. Whatever the reasons, the Hong Kong plan was never implemented.

Norway

The experience of Norwegian cities is worth mentioning because, even though the pricing is not based on congestion, the technology being used in two cities to collect the revenues could be readily adapted to congestion pricing. The cities of Bergen (1986), Oslo (1990), and Trondheim (1991) have put in place "toll rings" around each city. The cities decided to adopt toll collection as a means of raising funds for needed road improvements.[3] Toll booths or coin machines were placed at all the entry points around these cities, and a fee was imposed for entry. The fees are fairly low, ranging from U.S.$0.70 to $1.75, because the intent was not to change motorist behavior but to raise funds from the maximum number of users. Regular users can buy subscriptions that offer considerable discounts.

Oslo and Trondheim have implemented electronic road pricing.[4] A subscriber places a small box on the windshield behind the rear-view

[3] Some cities' decisions were influenced by a promise from the national government that it would increase its share of road financing for cities that would use toll financing (Gomez-Ibanez and Small forthcoming).

[4] Trondheim's toll pricing is close to congestion pricing since the city does impose a small (U.S.$0.32) premium over the normal toll during peak periods. The increment is not designed to reduce congestion, however, and is therefore not considered to be congestion pricing.

mirror. The box contains an electronic "tag" encoded with the motorist's identification. Passage over the entry point is recorded and the user is subsequently billed. Enforcement is handled by taking a video image of the license plates of vehicles that pass through the electronic gate without a tag, but evasion is rare. Although most users are subscribers, non-subscribers and irregular users can pay manually. About 80 percent of the revenue earned in Trondheim is collected electronically, and the toll collection technology in both Oslo and Trondheim costs about 15 percent of the total revenue earned. Public opinion polls ran heavily against toll collection before it was implemented. Motorist opposition has subsequently weakened somewhat, but about 50 percent continue to oppose the toll rings in both Oslo and Trondheim. In Trondheim, when motorists were asked whether they supported the toll ring given the improvements in road and public transportation facilities that it helps fund, only 25 percent said that they opposed it (Gomez-Ibanez and Small forthcoming).

Other Nations

Congestion pricing is being debated and studied in several Scandinavian and British cities. Political leaders in Stockholm have agreed to a toll ring proposal that would charge an average of U.S.$2.55 (in 1992 prices) to enter the central city or its inner ring road (Gomez-Ibanez and Small forthcoming). The proposal has not yet been adopted in Parliament, however, and many important details remain to be worked out, including whether the price will vary by time of day. Similarly, a proposal is being developed in the Netherlands for the Randstad region that would impose a charge for entering the primary road system during the morning peak. The Randstad is a polycentric metropolitan area of about 5800 km^2 (2,200 mi^2); it includes the nation's largest cities, airport, and port and is home to over 6 million people. An early version of the proposal would have imposed a multiple cordon pricing scheme with 140 collection sites throughout the region. The proposal now in Parliament is less ambitious, but still retains a peak-period increment (Gomez-Ibanez and Small forthcoming).

Congestion pricing is being actively considered in London and Cambridge, England. A 3-year, $5-million study is currently under way for London to examine alternative area pricing concepts and study public and political acceptance of the various proposals. The study is considering several pricing schemes, including one that would charge a higher price for operating in the most congested central area of London and a lower price for operating in the less congested area around the core (Gomez-Ibanez and

Small forthcoming; May, Vol. 2). The London study is expected to be completed in 1994. Congestion pricing has been studied for London several times in the past, but without leading to action (Button 1982).

U.S. Proposals

Proposals for congestion pricing are moving forward in the San Francisco Bay Area and in Southern California. In contrast to the hostile reaction to congestion pricing by local business leaders and elected officials 20 years ago, in a few places they are now publicly advocating this scheme.

The San Francisco Bay Area proposes to increase the existing $1 toll for using the San Francisco–Oakland Bay Bridge during the peak period by $1 or perhaps $2 (Dittmar, Vol. 2). The proposal was issued jointly by the Bay Area Congestion Pricing Task Force, which includes government, environmental, and business organizations. The intent is to shift some demand on the Bay Bridge to the off peak and to transit; this would reduce congestion and improve air quality. This proposal is the only congestion pricing pilot project that has been authorized under the Intermodal Surface Transportation Efficiency Act of 1991 (ISTEA).

On the Riverside Freeway (SR 91) in Southern California, a private operator will impose peak-period fees to allow solo drivers to use 16 km (10 mi) of newly constructed express lanes in the median of the existing highway (see Figure 2-3) (Fielding, Vol. 2). Construction of these lanes began in July 1993. Drivers who choose not to use the toll road can continue to drive on the existing road, which is heavily congested during the peak. Carpools with three or more occupants will be able to use the new lanes without charge. One other private toll road being planned in California, Route 57, will also use congestion pricing (Fielding, Vol. 2).

In the San Diego area, it has been proposed that solo drivers pay a toll to use an existing, underused HOV lane in the Interstate 15 corridor north of the city (Duve, Vol. 2). This proposal, along with several similar ones, was not selected by the interagency review panel established to review proposals for congestion pricing pilot projects. Members of the panel were apparently concerned that opening up HOV lanes, even for a price, would weaken the incentives for carpools. There was also concern that the proposal did not constitute congestion pricing because it was not pricing a congested facility, but rather pricing an underused (but restricted-use) facility to ease congestion on an unrestricted (but heavily congested) free route. Regarding the latter concern, the study committee considers the San

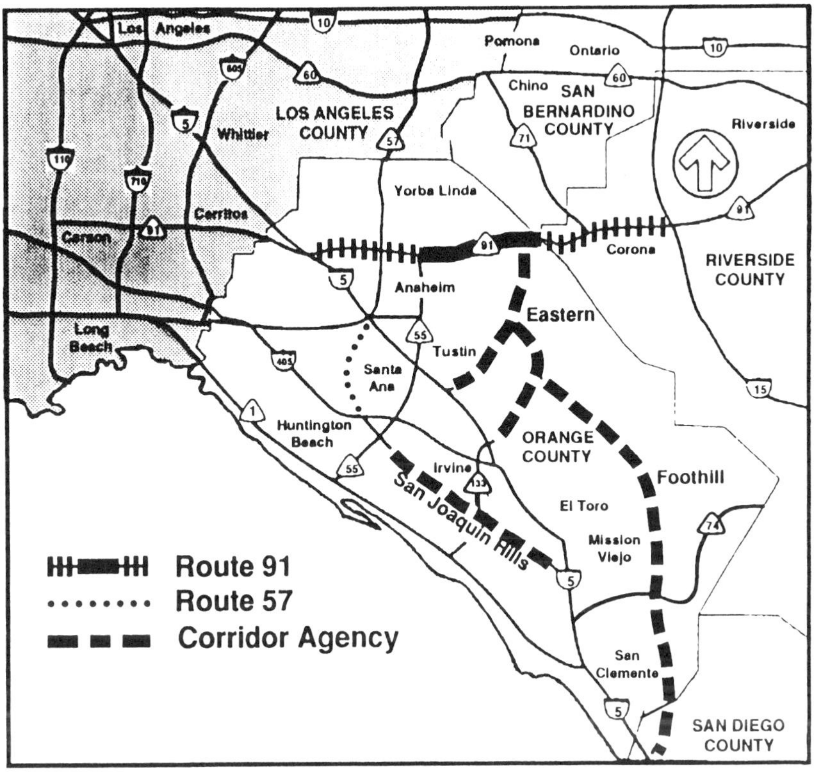

FIGURE 2-3 Route 91 and Route 57 projects are designed to use electronic congestion pricing.

Diego proposal to constitute congestion pricing because the end result is that motorists pay a fee to use these faster lanes during the peak hours. More specifically, it is congestion pricing with a carpool exemption and with an immediately adjacent, parallel free route. The proposal has the advantage of not taking anything away from motorists—instead it provides them with the option of paying for a premium service. The disadvantage with this proposal is that it could weaken a regional strategy to promote ridesharing on the HOV lane. Some motorists who formerly joined carpools in order to save time may elect to drive alone and pay for the time savings. San Diego could not implement this proposal on its own unless it was accepted as a pilot project because under ISTEA only pilot projects are able to introduce congestion tolls on Interstate routes.

Parking pricing strategies are another mechanism for imposing a higher fee for travel within or to a congested area. Parking charges do not deal with through traffic, which in some areas is the most significant source of congestion. Parking charges, however, can be designed to influence the choices of solo drivers. The "cashing out" of employee parking that has been adopted in California law requires employers who lease parking from a third party to provide their employees with the option of taking the cash equivalent of the employee parking benefit (Shoup, Vol. 2). Drawing on case studies of individual companies that changed their parking fees, Shoup estimates that if offered in the greater Los Angeles area, the cash-out option would reduce the number of solo driver trips to work by 20 percent (Shoup, Vol. 2, Table 5).

The California law was written to minimize adverse impacts on employers. Those that lease parking from third parties in a large block, and are therefore unable to return a few spaces, are exempt from the parking cash-out requirement until the term of the lease expires. Once the lease has been renewed, the employer must cash out parking to employees who prefer this option. Under this circumstance, the employer who wishes to minimize costs will need to negotiate with the lessor to reduce the number of spaces being leased. The law has another provision that requires cities and counties to reduce the parking requirements in zoning regulations for new and existing structures. This provision is designed to give lessors an incentive to convert land devoted to parking to higher uses. The ways in which the California law is designed to create benefits for employees, employers, and the general public are discussed in detail in Shoup's paper in Volume 2.

REFERENCES

Borins, S. 1988. Electronic Road Pricing: An Idea Whose Time May Never Come. *Transportation Research*, Vol. 22A, No. 1, Jan., pp. 37–44.

Button, K. 1982. Road Pricing: Some Practical and Economic Considerations for Its Introduction in Britain. *Logistics and Transportation Review*, Vol. 18, No. 4, pp. 353–371.

Dawson, H., and I. Catling. 1986. Electronic Road Pricing in Hong Kong. *Transportation Research*, Vol. 20A, No. 2, pp. 129–134.

Gomez-Ibanez, J., and K. Small. Forthcoming. *NCHRP Synthesis of Highway Practice: Road Pricing for Congestion Management: A Survey of International Practice*. TRB, National Research Council, Washington, D.C.

Hau, T. 1992. *Congestion Charging Mechanisms: An Evaluation of Current Practice*. Policy Research Working Papers. WPS 1071. The World Bank, Washington, D.C.

Wohl, M., and C. Hendrickson. 1984. Some Practical Pricing Problems. In *Transportation Investment and Pricing Principles*, Chap. 13, John Wiley and Sons, New York.

3
Possible Effects

Adopting congestion pricing would represent a significant change in operation of the transportation system, one in which there would be significant net benefits, but some losers. Given that congestion pricing has to be authorized in a political context, estimates of the relative sizes of the benefits and costs and identification of winners and losers will affect the quality and the outcomes of the debates that will occur.

Reviewed in the first section of this chapter are analyses of recent efforts to estimate the effects that congestion pricing might have on aggregate travel choices and how these changes would improve traffic flow in specific metropolitan areas. The second section provides estimates of the net economic benefits. The improvements in traffic flow for the majority of travelers would result from the choices about how, when, and whether to travel made by other travelers, especially motorists with low values of time or with flexibility in their schedules. Because of the anticipated impact on low-income motorists, more of the debate about congestion pricing in the United States has focused on the possible equity consequences of this policy than on its congestion reduction potential. This issue is taken up in the third section.

Business logistics and commercial transportation within metropolitan areas should benefit from congestion pricing, but resistance to higher prices by commercial interests has been a significant source of political opposition to toll increases and congestion pricing proposals in the past. These issues are reviewed in the fourth section.

Discussion about congestion pricing in the United States would not have advanced as far as it has without the emphasis on reducing pollution from mobile sources in the Clean Air Act 1990 Amendments (CAAA) and the growing interest in congestion pricing by environmental groups, business interests, and local and political leaders seeking market-based approaches to reducing congestion rather than regulations restricting travel. Plausible estimates of air quality and energy benefits are discussed next, followed by a review of the debate about how congestion pricing might affect the form of urban development and how this, in turn, might affect those firms located in areas or on routes subject to pricing.

Although this chapter reviews available studies, in the absence of any direct experience with congestion pricing in the United States, it is difficult to predict with complete confidence how changes in travel might unfold and how these changes in turn would affect specific groups, air quality, energy consumption, and urban form. There is sufficient experience with traveler responses to toll increases, transit fare changes, and changes in parking pricing to predict the direction and magnitude of changes in total travel within a metropolitan area, but estimating these consequences across different income groups and classes of highway users is more speculative. Available estimates of changes in emissions are also subject to considerable uncertainty. Longer-term consequences for metropolitan development are even more difficult to gauge. It is possible to estimate near-term consequences with models that forecast aggregate changes in travel, but it is important to appreciate the approximate nature of these estimates.

IMPACT ON TRAVEL

The evidence from past changes in bridge and turnpike tolls, transit fares, and parking fees demonstrates that motorists do respond to changes in price, even for commuting trips (Harvey, Vol. 2). Economists refer to the sensitivity in demand for a good that results from changes in price as the "elasticity" of a response. Elasticities are calculated in percentage terms: a price elasticity of demand of −2.0 would indicate that a 1 percent increase in price would result in a 2 percent decline in demand. When consumer sensitivity to a price increase lies between −1.0 and 0, demand is referred to as "inelastic." Thus, if the price elasticity is −0.5, a 1 percent increase in price will result in a 0.5 percent decline in demand. In this case, there will

POSSIBLE CHANGES IN TRAVEL CHOICES

Route Choice: to untolled routes or to tolled routes with shorter travel times.

Time of Travel: to earlier or later departures to avoid tolls or to tolled period for time savings.

Mode: to or away from carpools, transit, or other mode.

Destination: for nonwork trips, to shorter trips; for work trips, to changes in work or residential location.

Linked Trips: to more combination of errands on a single trip.

Trip Frequency and Activity Selection: for nonwork trips, to fewer discretionary trips; for work trips, to more telecommuting.

Automobile Ownership: to forgo ownership.

be a decline in demand following a price increase, but consumer demand for the product is strong enough that the percentage decline in demand will not be as large as the percentage increase in price. Analyses of individual behavior in response to price changes in transportation suggest an elasticity of "about −0.10 to −0.15 at the low end to −0.3 to −0.4 at the high end depending on the charge, the current costs of travel, and the capacity of alternative roads and transit systems" (Bhatt, Vol. 2). In other words, a 10 percent increase in price would result in a 1 to 4 percent decline in demand.[1] Although this response is inelastic compared with some other goods and services in the economy, a price increase does result in a decline in demand. Some travelers will change their behavior, and as illustrated in the previous chapter, when facilities are congested, the changes of a small proportion of motorists result in a more than proportionate improvement in traffic flow.

The range of possible behavioral adaptations, however, is quite complex (see text box). A large body of literature is available to inform estimates of mode shifts between automobiles and transit, but less support is available

[1] The price elasticity of demand for road use will vary substantially during the course of the day for a typical unit of roadway. Demand may be inelastic during the normal commute hours but elastic in mid-morning or mid-afternoon.

for estimates of shifts among single vehicle occupancy, carpools, and transit (Kain, Vol. 2). There is also relatively little empirical information on adjustments in the timing of trips, the frequency of linked trips, or changes in employment or residential location that might occur because of pricing. A major unresolved issue in the context of the effect of congestion pricing is how a price that varies with the level of congestion (and hence with the time of day) will affect the temporal characteristics of travel behavior. (This issue is discussed in more detail in the paper by Harvey that appears in Volume 2.) Some of the possible behavioral changes outlined in the text box are probably more important than others, but the "theoretical literature is not much help in sorting out first-order effects" (Harvey, Vol. 2), and empirical evidence indicating the magnitude of effects is quite sparse.

The most detailed and recent modeling of congestion pricing for specific U.S. metropolitan areas was applied to the San Francisco Bay Area and the South Coast Air Basin (the greater Los Angeles area) (Harvey, Vol. 2). In these analyses, the goal was to improve and maintain traffic flow on highways in the range of 48 to 72 km/hr (30 to 45 mph) by imposing congestion fees on all thoroughfares in the region during peak travel periods. A similar level of service was aimed for in both studies, but the Bay Area study uses travel and emission levels for 1997 as a basis for comparison and the South Coast Air Basin study uses 2010 travel and mobile source emissions estimates as a base line. To accomplish this level of service, the model results suggest that congestion fees averaging $0.06/km ($0.10/mi) would be needed in the San Francisco Bay Area and $0.09/km ($0.15/mi) in the greater Los Angeles area during peak periods. The fees are expressed as averages. On some facilities they would need to be considerably higher—as much as $0.37/km ($0.60/mi) on the most severely congested facilities. These results are roughly consistent with estimates developed from other approaches. (See Appendix B for results from other studies.) For an average round trip of 32 km (20 mi), these fees would result in daily round-trip costs of roughly $2.00 in the San Francisco Bay Area and $3.00 in Los Angeles. Fees of this magnitude would generate substantial revenues—about $3 billion annually in greater Los Angeles (Small 1992).

Regionwide congestion pricing would reduce total annual vehicle kilometers traveled (VKT) by 1.8 percent in San Francisco (Table 3-1) and 5 percent in Los Angeles (Table 3-2). These estimates are reductions in overall travel. Shifts in peak-period travel to the off peak would be considerably larger. For example, assuming that about one-fourth to one-third of

TABLE 3-1 Overview of the Bay Area Pricing Study (Harvey, Vol. 2)

		Percent Change from 1997 Mobile Source Baseline						
Strategy	Description	VKT	Trips	Fuel	ROG	CO	NO_x	CO_2
Regionwide congestion pricing (LOS D/E), average $0.10/mi	Automatic Vehicle Identification (AVI) scheme to price regional freeway and arterial system to maintain LOS D/E	−1.8	−2.2	−6.5	−5.5	−7.5	−2.9	−6.5
Regionwide employee parking charge, $3.00/day	All workers in region to experience minimum $3.00/day (1991) charge for parking automobile, pickup, or van at workplace	−1.2	−1.5	−1.2	−1.4	−1.4	−1.5	−1.2

NOTE: VKT = automobile and private transit vehicle kilometers traveled; Trips = automobile vehicle trips; Fuel = gallons of fuel consumed; ROG = emissions of reactive organics; CO = emissions of carbon monoxide; NO_x = emissions of oxides of nitrogen; CO_2 = emissions of carbon dioxide. LOS = level of service (a measure of traffic flow; see Appendix A). 1 mi = 1.6 km.

The accuracy of these estimates depends on uncertainties that are inherent in any travel forecasting exercise, such as in regional and subregional growth projections and assumptions about future infrastructure investments.

TABLE 3-2 Overview of the South Coast Air Basin Pricing Study (Harvey, Vol. 2)

		Percent Change from 2010 Mobile Source Baseline						
Strategy	Description	VKT	Trips	Fuel	ROG	CO	NO_x	CO_2
Regionwide congestion pricing (LOS D/E), average $0.15/mi	Automatic Vehicle Identification (AVI) scheme to price the regional freeway and arterial system to maintain LOS D/E	−5.0 [±.7]	−3.8 [±.4]	−9.2 [±.8]	−8.2 [±.8]	−12.1 [±1.0]	−8.4 [±1.0]	−9.2 [±.9]
Regionwide employee parking charge, $3.00/day	All workers in region to experience minimum $3.00/day (1991) charge for parking automobile, pickup, or van at workplace	−1.5 [±.2]	−1.8 [±.3]	−1.7 [±.3]	−1.7 [±.3]	−2.1 [±.3]	−1.6 [±.3]	−1.7 [±.3]

NOTE: VKT = automobile and private transit vehicle kilometers traveled; Trips = automobile vehicle trips; Fuel = gallons of fuel consumed; ROG = emissions of reactive organics; CO = emissions of carbon monoxide; NO_x = emissions of oxides of nitrogen; CO_2 = emissions of carbon dioxide. 1 mi = 1.6 km.

Each value represents the midpoint of the estimated range of effect. Numbers in brackets indicate variation above and below the midpoint based on sensitivity tests of key parameters related to pricing (such as the travel cost coefficients). Accuracy of the estimates will also depend on other uncertainties that are inherent in any travel forecasting exercise, such as in regional and subregional growth projections and in assumptions about future infrastructure investments.

trips occur during the peak, a 5 percent reduction in total travel would imply a 15 to 20 percent reduction in peak-period travel. This reduction in travel would result in time savings for the average congested peak-period round trip of about 10 to 15 min. As an alternative policy, regionwide parking charges on all employees of $3.00/day would reduce VKT by 1.2 percent in San Francisco and 1.5 percent in Los Angeles. Parking pricing would have a smaller effect on total travel in a congested area because it does not affect through traffic (Tables 3-1 and 3-2).

The 2 percent reduction in total travel as a result of congestion pricing fees of $0.06/km ($0.10/mi) in the San Francisco Bay Area may seem modest, but when compared with other, nonpricing travel demand management policies, it appears more substantial. By way of comparison, Deakin (1993) estimates that if the Bay Area implemented all reasonably available transportation control measures—including an employer-based trip reduction rule, improved transit services, reduced transit fares, and construction of carpool lanes—these measures combined might have roughly the same impact as congestion pricing on regional travel and emissions, but would impose a substantially higher cost on the region's economy.

NET BENEFITS

Congestion pricing on highways would have broad effects on the entire transportation system by shifting the demand for transportation services away from peak-period highway use by solo drivers. A reduction in the incentives to drive during peak periods would shift some traffic to the off peak, which would increase the efficiency with which the road system is used and reduce the demand for additional capacity. Some motorists would continue to drive during the peak but would elect to share rides with others or change the destinations of their trips. Sharing rides with others would also increase the efficiency with which the system is used by increasing the number of people per vehicle during peak periods.

Some motorists would shift to transit. The improvement in traffic flows that would result from congestion pricing would improve service reliability and speed, which would combine with the direct monetary incentive to make transit considerably more attractive, relative to the automobile, than it is today. The increase in ridership would improve system revenues (Kain, Vol. 2). The revenues could be used to expand service frequency or route coverage, which would make transit service all the more attractive to users (Kain, Vol. 2). With buses less delayed by traffic congestion, both drivers and buses could be used more productively.

Congestion pricing would reduce the demand for new highway capacity; this reduced demand would ease the capital requirements for expanding highways in response to growing population and travel demand. The U.S. Department of Transportation currently estimates that over the next 20 years the capital demands for the nation's roads and highways will range between $758 and $1,009 billion (in 1991 dollars) (DOT 1993). The low end of the range in costs represents the mere addition of sufficient capacity to maintain congestion at existing levels. The high end of the range represents the cost of easing congestion above existing conditions. The cost of expanding urban roadways in response to congestion will range between $232 and $365 billion. On an annual basis, the need for roadway expansion in urban areas totals $11.6 to $18.25 billion (of a total annual cost for urban and rural investment of $38 to $50.5 billion). Because of the phenomenon of latent demand, there is little prospect that these investments would actually reduce congestion. If congestion pricing were to be adopted nationwide, it would greatly reduce most of the new capital investment needs in urban areas (Winston and Bosworth 1992). It would result in time savings and reduced investment needs valued at roughly $5 to $11 billion (Small et al. 1989; Repetto et al. 1992).

DISTRIBUTIONAL EFFECTS

Similar to any other marketlike mechanism for allocating demand, the fees imposed by congestion pricing would cost a larger proportion of the income of low-income motorists, and presumably these motorists would be among the most likely to shift travel time or mode to avoid the fee. Because of the substantial revenues that congestion pricing can raise, however, it is possible to provide compensation to those groups most disadvantaged by congestion pricing. For example, to offset adverse impacts on the poor, some of the revenues could be used to reduce the regressive taxes (property, sales, and gasoline) used to finance roads and transit.

The revenues earned from congestion pricing could be used to ensure that all income groups benefit; this is important, because without the benefits that these revenues provide, the average commuter would be somewhat worse off than before (Small 1983; Hau 1992). The time gains of congestion pricing will not be as large as the out-of-pocket expense for the average commuter; thus the average commuter will benefit only if some of the revenues are returned to users in some form. (The form for returning the revenues could include anything from a property tax rebate to investments in new capacity or social investments approved by motorists. This issue is taken up in Chapter 4.)

The distribution of losses and gains across income groups can be illustrated by estimating the changes in commuting times and mode choices and the resulting time savings and losses, and by discussing how the revenues could be used to offset the disadvantages borne by certain groups. Small (1983) used the San Francisco Bay Area as a case study, analyzing a hypothetical $1.00 average peak-period expressway toll to reduce congestion. After valuing the time gains and losses, Small showed that before accounting for the revenue uses, the average lower-income commuter would "lose" $0.28 per day, the average middle-income user would "lose" $0.13 per day, and the average higher-income commuter would "gain" $0.08 per day (Small 1983).[2] The losses to the lower- and middle-income groups, however, can be completely reversed by specific uses of the revenues. For example, if the revenues earned were simply divided among all the individuals in a jurisdiction, all income groups would gain, but the lower-income group would still be disadvantaged. More progressive reallocations, such as using the revenues to offset regressive gasoline, property, and sales taxes, would benefit the lower-income groups more and higher-income groups less (Small 1983; Small et al. 1989). (The political feasibility of managing a revenue distribution system of this kind and its potential complexity are discussed in Chapter 4.)

Although the revenues earned could be used to compensate all disadvantaged groups, some individuals within these groups would still be disadvantaged. Congestion pricing induces some individuals to change their behavior. Individuals with the lowest value of time are the most likely to change the timing of their trips, join a carpool, or shift to transit. The availability of alternatives softens the impact, but they may still be at a disadvantage. Workers with less flexible schedules may be less able to shift their commute times to avoid the fee. Working women typically have more domestic and child-care responsibilities than working men and thus

[2] Kain (Vol. 2) suggests that Small underestimates the benefits of congestion pricing to existing transit users, many of whom have low incomes, by not estimating the benefits from improved transit operating speeds and service as a result of improved traffic flow.

are more likely to drive to work alone during peak travel periods (Rosenbloom and Burns 1994). Working parents of small children, regardless of job classification, typically are less able than other workers to adjust their schedules to avoid traffic congestion (Giuliano, Vol. 2). Lower-income working single parents, usually women, who rely on automobiles to get to work may be among those most directly affected. Despite the increased out-of-pocket expense, some working parents may appreciate the benefit of saving a few minutes in a hectic morning of getting children to daycare and trying to arrive at work on time, even if this cost does represent a larger share of their income. For those with no money to spare and little choice as to the mode or time of travel, however, congestion pricing could represent a substantial hardship.

Commuters with long trips to their jobs may not have many other options (aside from changing jobs) than to drive alone. Such individuals are likely to be net losers, even after the revenues have been reallocated (Giuliano, Vol. 2). (Chapter 4 contains discussions of how the revenues might be distributed to compensate groups adversely affected by congestion pricing and how the political feasibility of this concept could depend upon the vigor with which adversely affected individuals and groups might oppose it.)

A complete accounting of the potential winners and losers with congestion pricing also requires an assessment of how those who do not use automobiles could be affected (Gomez-Ibanez 1992). Current transit riders, among whom lower-income groups are more heavily represented, might experience more crowding if a significant proportion of road users shifts to public transportation. However, it is also plausible that the improvements in traffic flow, combined with increased revenues to transit agencies and followed by improved transit services, would result in net gains for transit users and transit agencies (Kain, Vol. 2). Research on the potential effects of congestion pricing on transit services and the benefits of improved services to lower-income users is of high priority. (This research need is discussed in Chapter 5.)

COMMERCIAL TRAFFIC

Estimates of the benefits of congestion pricing suggest that average speeds on arterial routes in Southern California could be improved by 8 to 16 km/hr (5 to 10 mph) and could be maintained at about 48 to 72 km/hr (30 to 45 mph). Although these changes appear small, they would make a substantial improvement in traffic flow. On major facilities, speeds of 40 km/hr (25 mph) or less imply that congestion is so heavy that minor interruptions can lead to complete breakdowns in flow. Under such conditions the probability of minor traffic accidents increases, and such incidents can cause long delays. Roughly half the delays are attributed to incidents that disrupt traffic flows (Schrank et al. 1993). Thus, maintaining a steady flow, even at modest speeds, can substantially improve traffic conditions.

Faster-moving traffic and more reliable traffic conditions would be valuable to commercial road users in congested metropolitan areas. A wide variety of commercial users depend on the road system. Although the large trucks delivering cargo to stores and factories come immediately to mind, commercial users also include panel trucks carrying package freight to stores and homes and light-duty vehicles driven by carpenters, electricians, plumbers, and delivery workers. Salespeople are also regular users of the road system, as are business people traveling between appointments or making trips to airports. Congestion causes immediate productivity losses to employers whose workers are stuck in traffic. Delays in the delivery of goods mean that physical assets are not being used efficiently. Increases in the transit time of cargo increase inventory costs. Although commercial users are among those highway users least able to shift to avoid congestion fees, it would appear that they are also among the users most likely to gain. Studies or analyses of the assumed benefits of congestion pricing to business activity and commercial transportation, however, have not been made for U.S. metropolitan areas.

AIR QUALITY

Concern about meeting state and federal air quality standards is a prime motivator behind the renewed interest in congestion pricing. Despite considerable reductions in several types of emissions over the last 20 years, many metropolitan areas are still unable to meet air quality standards for ozone and carbon monoxide. Emissions from automobiles are major sources of these pollutants. Some 94 urban areas are currently out of compliance with federal standards for ozone (EPA 1993); the 1990 CAAA requires those areas classified with "moderate" or higher nonattainment (55 urban areas) to develop specific measures to reduce the emissions caused by automobiles. EPA has designated 42 areas as being in nonattainment for carbon monoxide.

The majority of reduced air pollution from mobile sources will come—as it has in the past—from policies affecting vehicles. Such policies have included tailpipe emission standards, regulations on fuel content and mixture, and improved vehicle inspection and maintenance programs. In some urban areas, however, these will not be enough to attain compliance with federal (and California) standards. In those areas, there is no choice but to try to change motorist behavior. Efforts to change behavior through specific policies, such as the requirement that firms with more than 100 employees develop programs to reduce the occupancy of the vehicles their employees drive to work, have yielded modest reductions in emissions but have done so at a high cost relative to the benefits (Giuliano and Wachs 1992). Congestion pricing would provide a more cost-effective approach to achieving the required reductions.

The Bay Area Economic Forum (1990), a public-private partnership of governments and business in the San Francisco Bay Area, states, "The [California] Clean Air Act requires reducing vehicle emissions by one-third. We face tough decisions." The Bay Area is required under California's clean air law to adopt transportation control measures that will reduce total trips, vehicle miles traveled, and traffic congestion. The Forum has strongly advocated that the Bay Area adopt market-based approaches to meeting air quality goals rather than relying on regulation, and has succeeded in garnering the support of local governments, businesses, and some in the media.

In modeling the effects of congestion pricing for the Bay Area and the greater Los Angeles area, Harvey estimates that regionwide congestion pricing could result in reductions in two pollutants that are precursors to the formation of ozone—oxides of nitrogen and reactive organic compounds (which include hydrocarbons). Congestion pricing for the Bay Area and Los Angeles is estimated to reduce reactive organics by 5.5 to 8.2 percent and oxides of nitrogen by 2.9 to 8.4 percent. These calculations of emission benefits, however, rely on assumptions about emissions reductions that involve a number of uncertainties.

Reductions in regional ozone and other emissions that might result from congestion pricing are quite difficult to predict for a variety of reasons.

- Estimates of the relationships between the emission rates of vehicles on the road and changes in average speeds are uncertain because of weaknesses in the data used to calibrate existing models (Guensler and Sperling, Vol. 2).

• Reductions in congestion, or improvements in traffic flow, will reduce some emissions, such as reactive organics and carbon monoxide, particularly if the previous average speed was below 32 km/hr (20 mph), but as speed increases, emissions of oxides of nitrogen, another precursor to ozone formation, increase as well (Guensler and Sperling, Vol. 2).

• Improvements in the flow on a corridor can also affect the traffic flows on adjacent corridors, the net result of which can either increase or decrease emissions on the improved corridor (Horowitz 1982, 193).

Assuming that in a regionwide congestion pricing program average speeds increase from less than 32 km/hr (20 mph) and do not increase to more than 64 km/hr (40 mph), emissions of all kinds are likely to be reduced (Guensler and Sperling, Vol. 2). The net effects that reductions in reactive organics and oxides of nitrogen would have on ozone formation within a region, however, will vary according to local atmospheric conditions. The formation of ozone in the lower atmosphere is very complex; oxides of nitrogen and reactive organic emissions mix to produce ozone at different levels depending on ultraviolet radiation, temperature, and the prevalence of other natural sources of nitrogen oxides (National Research Council 1991). In general, however, if increased speeds resulting from congestion pricing do not exceed 64 km/hr (40 mph), NO_x and volatile organic compounds (VOCs) would decline and ozone formation should lessen (Guensler and Sperling, Vol. 2).

As indicated in the paper by Guensler and Sperling that appears in Volume 2, there is a great deal of uncertainty about how vehicular emissions change with speed. Given this uncertainty, these authors have been cautious in defining a band of speeds in which emissions would be reduced [between 32 and 64 km/hr (20 to 40 mph)]. The California Air Resources Board and the Environmental Protection Agency have research under way that is designed to reduce the uncertainty about the relationship between emissions and vehicular speeds. The outcome of this research has important implications for the air quality benefits of congestion pricing. The emissions/speed curves in the paper by Guensler and Sperling indicate that emissions of hydrocarbons and carbon monoxide continue to fall between 64 and 80 km/hr (40 to 50 mph) and that even when they increase at higher speeds, they remain well below the emission levels that occur in congested conditions. Thus improvements in traffic flow resulting from congestion pricing should reduce these emissions. For oxides of nitrogen, however, speeds above 64 km/hr (40 mph) may actually increase the amount of these emissions compared with the amounts that would occur in congested conditions. In metropolitan regions that are primarily concerned about controlling oxides of nitrogen, this implies that a pricing strategy would have to be very finely tuned to allow speeds to increase above stop-and-go conditions, but not so high as to allow free-flow conditions. Such fine tuning may not be practical with simple peak/off-peak pricing. Depending on how ongoing research reduces the uncertainty about the speed/emissions relationship and on the concern about oxides of nitrogen as opposed to other emissions, this could have implications for political feasibility by weakening the potential support of environmental groups.

Congestion pricing could reduce some ozone precursor emissions by simply reducing the actual number of trips taken. A large share of such emissions is caused by starting cold vehicles; hence a reduction in trips would reduce these emissions directly. About 50 percent of hydrocarbon emissions (reactive organics) result from cold starts and another 10 percent from evaporative emissions (the evaporation of emissions when the vehicle is not running but the engine is still hot).

In contrast to regionwide congestion pricing, estimates of emissions reductions from pricing individual facilities such as bridges or tunnels are highly dependent on the availability and capacity of alternate routes and on how much traffic would shift (Horowitz 1982). The availability of alternate routes may increase or decrease net emissions of carbon monoxide or oxides of nitrogen. To illustrate, consider two routes that serve the same origin and destination and that are reasonable substitutes for one another. Traffic on the two routes will tend to balance out as users shift back and forth to minimize trip times. If one of these routes becomes congested, some motorists will shift to the other. Alternatively, if one route were to be priced, demand for this route could be reduced during the peak and some traffic could shift during the peak to the alternate, unpriced route. Smoothing the traffic flow would reduce emissions of reactive organics or carbon monoxide for the priced route, but they could become worse on the alternate route if it became congested. These potential traffic shifts highlight the importance of designing congestion pricing pilot programs to avoid creating congestion at formerly uncongested points.

ENERGY

Regionwide congestion pricing would reduce energy consumption more directly than emissions. Fuel use would decline from both reduced trips and improved traffic flow. Harvey (Vol. 2) estimates a 6.5 percent reduction in fuel consumption in the Bay Area as a result of an average toll of \$0.06/km (\$0.10/mi) and a 9.2 percent reduction in the South Coast Air Basin as a result of an average toll of \$0.09/km (\$0.15/mi) (Tables 3-1 and 3-2).

URBAN FORM

Over the long term, changes in land use patterns could have substantial impacts on air quality and energy in addition to those already cited. Two different arguments are made about the possible effects of congestion pricing on the spatial development of metropolitan areas (Deakin, Vol. 2). One line of reasoning holds that the traditional underpricing of highways has encouraged urban sprawl and that correct pricing would encourage more dense development in and around urban centers. An opposing line of reasoning holds that congestion pricing would facilitate continued decentralization because *(a)* it would reduce the attractiveness of areas affected by pricing by creating a negative image (especially if competing commercial areas are unpriced) and *(b)* zoning regulation would prevent landholders from increasing the density of development.

The outcome of this debate has both practical and political implications. At the practical level it is important to anticipate the effects on development in order to be able to design a program that would minimize adverse consequences. At the level of political feasibility, it is important to anticipate the possible resistance of affected interests. When areawide pricing was proposed for the most congested parts of Manhattan in the mid-1980s, for example, retail and commercial establishments in the area reacted with alarm, apparently thinking that increased out-of-pocket costs for travel would reduce the attractiveness of the central business district.

Neither theory nor research on the relationship between the cost of transportation and urban development provides compelling evidence to support whether congestion pricing would have a centralizing or decentralizing effect (Deakin, Vol. 2). The possible impact of congestion pricing on urban form is complicated further by the intended use of the revenues and by steps that businesses might take to minimize the impact. For example, if the revenues are used to compensate adversely affected groups (which might include commercial establishments in an affected area), the land use impact might well be minimized. Alternatively, if the revenues were used to expand highway capacity at a major bottleneck serving the downtown area, it is possible that the improved access would benefit the affected area and facilitate residential development in outlying areas. Regarding strategies that businesses in affected areas might take to minimize the impact on their employees, they could shift hours of operation, encourage telecommuting, reimburse workers who pay tolls, or relocate some or all of their operations to less congested areas. The complexity of the potential responses to congestion pricing makes it all the more difficult to predict its effect on urban form.

As an alternative to affecting land use patterns, congestion pricing could affect the balance between the location of jobs and residences. In other words, congestion pricing could induce workers to move closer to their work sites. Such shifts could substantially reduce the total amount of travel required for work trips. Commuting times that exist within the current distribution of jobs and residences in the Los Angeles metropolitan area are two-thirds longer than what would be required if workers were located in neighborhoods that minimized their commutes (Small and Song 1992). Whether congestion pricing fees would induce substantial changes in residential location, however, would depend on the strength of other influences on residential location choice. The extra commuting that occurs already may be explained by preferences for neighborhood amenities such as schools or low crime rates, the difficulty of minimizing commutes for both workers in dual-worker households, and other influences, such as racial discrimination (Giuliano and Small 1993).

SUMMARY

The experience with pricing transportation in various modes provides a firm basis for predicting that congestion pricing would change traveler behavior and reduce congestion. The full range of potential behavioral responses is more complex than can be modeled with precision, but plausible estimates of the average motorist's response to price increases can be made. Congestion fees would cause some motorists to shift away from driving in the peak period. Estimates for the Bay Area and the Los Angeles area suggest that fees averaging $2.00 to $3.00/day on all thoroughfares in these regions would reduce peak-period travel by 10 to 15 percent and save average commuters 10 to 15 min per round trip.

Congestion pricing would have valuable systemwide effects. It would increase the efficiency with which existing assets are used by encouraging trip making in off-peak periods, ridesharing, and use of transit. If adopted in all metropolitan areas with traffic congestion problems, congestion pricing could result in net savings ranging from $5 to $10 billion.

Motorists with low incomes are among those most likely to be disadvantaged by congestion pricing. (Alternative means of compensating adversely affected groups are discussed in the next chapter.) Working women, particularly those with child-care responsibilities, are less able to adjust their schedules to avoid traveling during the peak period. They are therefore more likely to have to pay congestion fees. Women in higher income brackets may well value the time-saving benefit of less traffic congestion, even if it does cost more in out-of-pocket expenses. Lower-income working women with child-care responsibilities, particularly single parents, would be among those most disadvantaged. Some businesses in central areas and individuals living in geographic areas without workable alternative forms of transportation may also be among those adversely affected. Travelers with higher values of time stand to be the primary beneficiaries. In theory, commercial travelers would be among those with higher values of time and would therefore enjoy net benefits. Insufficient information is available, however, to draw conclusions about the net effect of congestion pricing on the poor, on businesses in areas affected by congestion pricing, and on commercial users. These are high-priority areas for future research.

Because congestion pricing would alter trip making and total travel, it would result in reduced air pollution from automobiles and would save energy. Regionwide congestion pricing in San Francisco and Los Angeles averaging $2.00 to $3.00/day, for example, could reduce emissions from automobiles by 3 to 9 percent, depending on the type of emission. Automotive fuel consumption would decline by 6.5 to 9.2 percent.

Plausible arguments can be made that congestion pricing would either increase or reduce the density of development within regions, but neither theory nor available research leads to a definitive conclusion in favor of either argument. The specific form of congestion pricing imposed, the reactions businesses might choose to mitigate its effects, and the uses of the revenues would all affect how land uses might change.

Past studies, including those drawn upon in this chapter, have assumed regionwide congestion pricing, but the proposals actually moving forward in the United States would price individual facilities. The Route 91 project in Orange County, the Bay Bridge proposal for San Francisco, and the other congestion pricing pilot project proposals received by the Federal Highway Administration concern individual congested routes or bridges. Because of the uncertainties about the actual impacts of congestion pricing and because of the institutional complexities and political difficulties of regional congestion pricing proposals (discussed in the next chapter), incremental changes in the direction of congestion pricing may be the only way that this policy would be implemented in the United States.

Incremental movements toward congestion pricing represented by current proposals, however, cannot be expected to have substantial effects on regional air quality, energy consumption, or land use. (The Bay Area proposal would be something of an exception, because the Bay Bridge carries such a large share of total traffic into San Francisco during the peak period.) Such effects may be small at the regional level, but could still have substantial local benefits. For carbon monoxide in particular, the health risks are caused by high concentrations, which result from heavy traffic flows, especially congested traffic. By smoothing traffic flow, congestion pricing on individual facilities can substantially reduce risks to health. In applying congestion pricing to individual facilities, however, it is important to anticipate traffic diversion to alternate routes and to design pilot programs such that new areas of congestion are not created.

Studies of the application of congestion pricing on the Bay Bridge, the Route 91 project, and on individual toll roads would not address all the uncertainties about this policy, but careful evaluation of driver responses in these situations would greatly improve the ability to estimate the consequences of congestion pricing in a broader context. Such evaluation could also delineate more clearly who wins and who loses (an evaluation design is outlined in Chapter 5).

REFERENCES

ABBREVIATIONS

DOT U.S. Department of Transportation
EPA Environmental Protection Agency

Bay Area Economic Forum. 1990. *Market Based Solutions to the Transportation Crisis: Incentives to Clear the Air and Ease Congestion.* San Francisco, Calif.

Deakin, E. 1993. Policy Responses in the U.S.A. In *Transport, the Environment, and Sustainable Development* (D. Banister and K. Button, eds.), E. and F.N. Spon, New York.

DOT. 1993. *The Status of the Nation's Highways, Bridges, and Transit: Conditions and Performance.* Report of the Secretary of Transportation to the U.S. Congress. Government Printing Office.

EPA. 1993. *National Air Quality Emissions Trends Report, 1992.* Report 454/R-93-031. Office of Air Quality Planning and Standards, Research Triangle Park, N.C.

Giuliano, G., and K. Small. Forthcoming. Is the Journey to Work Explained by Urban Structure? *Urban Studies.*

Giuliano, G., and M. Wachs. 1992. A Comparative Analysis of Regulatory and Market-Based Transportation Demand Management Strategies. In *Papers Presented at the Congestion Pricing Symposium, June 10–12.* FHWA and FTA, U.S. Department of Transportation.

Gomez-Ibanez, J. 1992. The Political Economy of Highway Tolls and Congestion Pricing. *Transportation Quarterly*, Vol. 46, No. 3, pp. 343–360. (For a summary, see Presentation Summary, In *Exploring the Role of Pricing as a Congestion Management Tool*, Searching for Solutions: A Policy Discussion Series, No. 1, U.S. Department of Transportation, March.)

Hau, T. 1992. *An Economic Analysis of Road Pricing: A Diagrammatic Approach.* Policy Research Working Papers. WPS 1070. The World Bank, Washington, D.C.

Horowitz, J. 1982. *Air Quality Analysis for Urban Transportation Planning.* The MIT Press, Cambridge, Mass.

National Research Council (Committee on Tropospheric Ozone Formation and Measurement). 1991. *Rethinking the Ozone Problem in Urban and Regional Air Pollution.* National Academy Press, Washington, D.C.

Repetto, R., et al. 1992. *Green Fees: How a Tax Shift Can Work for the Environment and the Economy.* World Resources Institute, Washington, D.C.

Rosenbloom, S., and E. Burns. 1994. Why Working Women Drive Alone: Implications for Travel Reduction Programs. To be published in *Transportation Research Record*, TRB, National Research Council, Washington, D.C.

Schrank, D., et al. 1993. *Estimates of Urban Roadway Congestion.* Research Report 1131-5. Texas Transportation Institute, College Station.

Small, K. 1983. The Incidence of Congestion Tolls on Urban Highways. *Journal of Urban Economics*, Vol. 13, pp. 90–111.

Small, K. 1992. Using the Revenues from Congestion Pricing. *Transportation*, Vol. 19, No. 4, pp. 359–381.

Small, K., and S. Song. 1992. "Wasteful" Commuting: A Resolution. *Journal of Political Economy*, Vol. 100, No. 4, pp. 888–898.

Small, K., C. Winston, and C. Evans. 1989. *Road Work: A New Highway Pricing and Investment Policy.* The Brookings Institution, Washington, D.C.

Winston, C., and B. Bosworth. 1992. Public Infrastructure. In *Setting Domestic Priorities* (H. Aaron and C. Schultze, eds.), The Brookings Institution, Washington, D.C.

4
Technical and Political Feasibility

The feasibility of congestion pricing will be affected by two types of issues: the technical ones and the political ones. Technical issues include the ability to charge users with minimal inconvenience and intrusion on privacy. Dealing with convenience and privacy is largely a matter of technology, and given the advances in this field, it is not an important barrier to implementation. Feasibility also requires having an institutional mechanism to administer the program. At first glance this probably does not seem like a major concern, but congestion in metropolitan areas often crosses local government and even state lines. The lack of existing institutions bridging city, suburban, county, and state government lines makes addressing regionwide congestion a vexing problem (Downs 1992). Addressed in the first section of this chapter are the availability of *(a)* technology to price efficiently and avoid privacy concerns and *(b)* institutional forms to resolve cross-border congestion.

Political feasibility is much more complex. Actually, it is a combination of issues, the most important of which are public acceptance of direct payment for road use, which is widely viewed as a free good, and the fairness to those unable to pay congestion fees without economic hardship. In the second section of this chapter, polling data on the public's willingness to pay to avoid congestion are reviewed, how revenues could be used to ameliorate the distributional impact of alternative congestion pricing programs is illustrated, and alternative strategies for building political support are outlined.

TECHNICAL FEASIBILITY

Available Technologies

In the few instances when congestion pricing was proposed in the past, policy makers eschewed it, in part because of the perception that having queues of motorists at toll booths to pay congestion prices would be self-defeating. Advances in electronic toll collection technology, however, are rapidly overcoming this problem (Pietrzyk, Vol. 2; May, Vol. 2). Some toll roads in the United States and abroad are already offering electronic toll collection services. A wide variety of technologies is already in use to charge users automatically as they drive by a collection point. Regular patrons of toll roads can purchase inexpensive electronic tags, which are read by a computer that identifies the vehicle and debits the account of the patron with each use. The tag can be permanently mounted on the license plate or under the car, or placed inside the car on the dashboard; some tags are about the size of a standard credit card and about twice as thick. These electronic transactions can occur at speeds of 88 km/hr (55 mph) or more without delaying motorists (Pietryzk, Vol. 2; May, Vol. 2). Vendors claim that their technologies are over 99 percent accurate; existing technology, which is not as advanced as that to be introduced, has an operating reliability in the range of 93 to 98 percent (Pietrzyk, Vol. 2).

Relatively simple tags are of a "read only" design, in which the toll road authority computer reads the identity of the passing vehicle and debits that account with each use. If a prepayment method is used, the patron purchases a certain number of uses in advance, and his or her account is reduced with each use. In a postpayment method, the individual is simply billed. Most users of these systems handle their monthly transactions with a credit card. Privacy can be protected by handling prepayment transactions in cash. A "read/write" tag allows the authority computer to identify the patron and to record transactions on the patron's tag, which gives the patron an independent record of transactions to check against the authority's billing. These tags are somewhat more sophisticated than those of the read-only design; they require a battery or a connection to the vehicle's battery. Purchases in cash to protect privacy would be somewhat inconvenient. In order to encourage the use of electronic toll collection, some toll road authorities will set up blind accounts in which billings are mailed to post-office addresses.

Toll road authorities have great incentives to move toward electronic toll technologies. Electronic toll collection allows for vehicular throughput of 1,200 to 1,800 vehicles per lane-hour compared with 350 to 500 vehicles per lane-hour with conventional attended or automatic toll booths. It also greatly reduces the cost of operation. Capital, operating (including collection and billing), and maintenance costs of electronic toll collection could be only 10 percent of the cost of operating conventional toll booths (Pietrzyk, Vol. 2).

A practical problem with automated systems is how to handle the occasional user or traveler from outside the area whose vehicle is not outfitted with the technology. Toll road authorities using automated systems continue to have attended toll booths. Some authorities separate these toll booths from the main traffic flow, which gives users with automatic technology free-flow conditions. Candidate routes and bridges for congestion pricing in urban areas, however, may not have space for such separate toll booths. Costs in these cases would be higher and traffic flow substantially lower if toll booths have to be provided for occasional users. Lack of space to provide an extra lane for occasional users may render some proposals for electronic pricing impractical.

Privacy

American motorists could be expected to resist toll collection technologies that invade their privacy. Privacy concerns were among the reasons that neighborhood councils in Hong Kong rejected congestion pricing in 1985 (Borins 1988). As mentioned above, however, toll road patrons on routes with electronic toll collection can now set up cash accounts that protect their privacy, and some toll authorities will set up billing accounts in which the identity of the patron is not revealed to the authority. In addition, the further development of technology will soon allow motorists to handle cash transactions through an expanded read/write technology that further protects motorist privacy. Read/write cards can be modified to become memory cards. With this technology, the patron purchases a card that stores the prepayments purchased. When the patron reaches a collection point, the memory card confirms to the roadside computer that sufficient credit is stored on the card for the patron to pass. With each passage, the computer deducts the toll from the card's value. With such a technology there is no need to create a record of when and where individuals traveled on a tolled facility.

Enforcement

Enforcement also involves privacy issues. Toll road authorities that collect tolls electronically use automatic enforcement, whereby photographs are taken of the license plates of vehicles that do not have transponders or active accounts (Pietryzk, Vol. 2). Most states, however, do not have legislation allowing such evidence to be used in court; those that do may also require that the infraction be witnessed by a policeman (Pietryzk, Vol. 2). Illinois was the first state to convict a driver of a traffic violation on the basis of photographic evidence. On the Dallas North Tollway, however, patrons soon learned that photographic evidence of toll road infractions would not stand up in court. Texas state police currently enforce such infractions as they would any other traffic violation. New York recently passed legislation requiring that toll road violations be witnessed by an "enforcement agent." Florida has a similar law, but recently passed legislation allowing photographs to also be used as evidence in court. Although the lack of automated enforcement increases costs, this appears to be an area in which law and practice are evolving to adapt to new technology. For example, the state of New York recently allowed the city of New York to use photographic evidence of speeding vehicles in issuing citations against the owners of these vehicles (except in cases where the vehicle had been stolen).

Administration

Traffic congestion in metropolitan areas is a problem that usually requires regional solutions (Downs 1992). Individual jurisdictions may find it difficult to solve traffic congestion without coordination with neighboring jurisdictions, and even if they manage to reduce congestion on their streets, they may simply shift it to an adjacent community. In most metropolitan areas, there is no appropriate institution for managing a regional congestion pricing or parking management program. Moreover, local governments typically oppose the formation of new regional entities that encroach on their traditional power and influence (Downs 1992).

Governmental structures within metropolitan areas are generally decentralized and fragmented. This fragmentation is not surprising. Americans have always distrusted centralized power and authority. In many metropolitan areas, the movement of people and jobs to the suburbs was quickly followed by the incorporation of entities separate from central cities, and these suburban communities resist encroachment by central city and metropolitan forms of government (Olson, Vol. 2).

Aside from the difficulties of creating a regional institution caused by intergovernmental competition, the appropriate form of a regional institution raises complex questions about the proper balance between independent authority and control versus accountability to the populace (Olson, Vol. 2). Such an institution should have sufficient authority and independence from political pressure to set and adjust prices at efficient levels and should be able to prevent the raiding of its treasury by general governmental units during periods of financial difficulties. Sufficient safeguards should be put in place to avoid a "gold-plated" organization with inflated salaries and high costs generally. More important, the institution would need to be sufficiently accountable to the public to ensure that it hews to its mission, but not so accountable that it would be unable to adjust the toll prices with changes in demand.

In his review of existing governmental units, Olson (Vol. 2) does not find an ideal form to recommend. Such an institution would have to meet several important criteria, some of which conflict. Accountability and efficiency, for example, are two governmental goals that routinely conflict. In addition, the tremendous variability in governmental structures across metropolitan areas precludes the identification of a single form that would work in all areas. Olson does, however, favor three options: state transportation departments, metropolitan districts, and special purpose local governments.

State transportation departments offer several advantages: they have sufficient authority, can operate on a statewide basis, have considerable professional expertise, and are accustomed to administering large sums of money. Like metropolitan districts, however, they are often perceived at the local level to be heavy handed and unresponsive to local concerns (Olson, Vol. 2). Because metropolitan areas are not directly represented within state transportation departments, as they would be in a metropolitan district, there may be insufficient trust at the local level to grant a state agency the power to operate a metropolitan-wide congestion pricing system.

Metropolitan districts are generally allowed under state enabling legislation. They are typically nonpartisan, cover an entire metropolitan area, and have broad authority to conduct their tasks. Transportation examples include transit agencies in Chicago, San Francisco, and elsewhere. State transportation departments and metropolitan districts are appealing because their spatial coverage suits them well to metropolitan-wide congestion problems.

Special purpose local governments (also referred to as special districts) are smaller-scale versions of metropolitan districts—with the same strengths and weaknesses. Because of their autonomy, they are less accountable than general purpose governments, which has led to some criticism about their being unresponsive to local concerns. Nonetheless, this institutional form may be the closest to an appropriate unit for managing regional congestion pricing that can be achieved (Downs 1992; Olson, Vol. 2). For congestion pricing projects limited in scope to a tunnel, bridge, or highway, an organization with less than a regionwide scope might well suffice (Olson, Vol. 2).

In response to budgetary shortfalls in the late 1980s, interest emerged in having the private sector build and operate roads (Gomez-Ibanez et al. 1991). Private intercity toll roads have become more common in Europe in recent years. In the United States, private toll roads are being developed in Northern Virginia, California, and Colorado, and are being considered in other states. Although Olson considers privatization of individual highway links, he dismisses privatization for metropolitan-wide applications because the private entity would be placed in the position of a monopoly supplier. Monopoly suppliers pervade the utility industry, however, and this problem is routinely handled through regulation by independent public bodies. Given that road networks are quite similar to networks of water pipes and networks of local telephone lines, consideration should be given to creating roadway utilities to implement congestion pricing (Roth 1991). Utilities can operate regionwide, routinely manage large sums of money, and could be created with adequate authority to carry out their tasks. The interests of the public could be protected in the same ways that they are protected with existing water, electric, gas, and telecommunications utilities.

Such a radical change in the organization of road authorities seems more than existing governmental agencies and the public could swallow in the short run, and having to create new utilities for something as controversial as congestion pricing may not help its feasibility. The regulated utility model may have some appeal over the longer term if congestion pricing gains acceptability with the public. Even so, the evolution of existing transportation entities into regional roadway utilities or the creation of regional roadway utilities would clearly not occur for a long time. The growth in privatization, both for individual facilities and broader networks of roads, involves many uncertainties about the potential efficiency gains, winners and losers, liability, and political feasibility. These would be fruitful topics for further research.

During the current period of small-scale experimentation, a variety of institutional forms is likely to emerge, each suited to the specific legal and political conditions in the areas where proposed (Olson, Vol. 2). For regional or areawide congestion pricing, it is clear that some form of regional institution is required because of the many interjurisdictional issues that are raised. Metropolitan special districts or state transportation agencies may work in some areas. The considerable variation in customs, administrative arrangements, and law from locale to locale almost dictates that different institutions emerge. Although no single institutional form appears perfect, there are sufficient existing models to suggest that satisfactory arrangements could be achieved.

POLITICAL FEASIBILITY

A shift from the current approach to highway finance and management to a congestion pricing approach would represent a significant—even radical—change. The political feasibility of achieving such a change is therefore difficult to assess. Three issues appear paramount in determining political feasibility: the general acceptance by the public of paying for facilities at the time and place used as opposed to paying in the form of taxes on gasoline, sales, and property; the distribution of winners and losers (which includes revenue uses designed to minimize the adverse impacts); and the conditions influencing local coalition building in support of or in opposition to congestion pricing.

General Acceptance

American motorists are accustomed to using roads without paying for use at the time. Although motorists would be expected to resist increased tolling of any kind, opinion polls suggest a willingness to pay by users confronted with congestion. A national opinion poll found that two-thirds of motorists who experience congestion say they would pay to avoid delay (Apogee Research, Inc. 1990). About 4 in 10 would pay at least $2.00/day and another 2 in 10 would pay at least $0.50. In San Francisco, public opinion surveys in 1988 and 1990 indicated that two-thirds of respondents supported a $1.00 increase to the existing toll on the Bay Bridge as an acceptable price to pay for cleaner air (Dittmar, Vol. 2; Bay Area Council, Inc. and KQED, Inc. 1991).

The openness to experimenting with congestion pricing in San Francisco and Southern California stems from several factors. Funds for road building are scarce because of constraints on the ability of the state to raise taxes to support capacity expansions. Significant air quality problems—particularly in the Los Angeles area—coupled with stringent air quality standards and strong neighborhood opposition to new roads limit the ability to expand highway capacity. Requirements in California law to restrict solo driving through regulation may be less appealing than congestion pricing.

An analogy about public support for congestion pricing can be drawn from survey research in England. Whereas 57 percent of English respondents responded negatively to a question that asked whether they would favor congestion pricing in the abstract, this opposition was reversed when respondents were asked whether they would support congestion pricing if the funds were used to improve public transportation, reduce accidents, and improve pedestrian and cycling facilities. In response to this question, 57 percent supported congestion pricing (Jones 1991). American voters can be expected to resist congestion pricing if it is perceived as merely another tax, but they may support the policy if other conditions hold, such as severe congestion, support for cleaner air, unpalatable alternatives, and dedication of the revenues to purposes they support. The use of the revenues is a critical feature of any proposal and is closely linked with general public acceptance. This element is discussed in more detail in the next section.

Public support is far from guaranteed, however, even when the funds are dedicated to transportation purposes. In Norway, for example, where toll rings are being used in three cities to raise revenues to provide new transportation capacity, a simple majority in support of the toll rings has been achieved only in Bergen (Gomez-Ibanez and Small forthcoming). Support in Oslo grew from 29 percent in 1989 to 39 percent in 1992, and those opposed dropped from 65 to 56 percent. When Trondheim residents were asked in 1991 whether they supported the toll rings given that the revenues would be used to provide specific facilities, opposition dropped from 48 to 23 percent, but those unsure about their support grew from 32 to 48 percent. Support increased from only 20 percent to 32 percent of respondents—a plurality in support, but not a majority.

Public opinion polls in the United States suggest a receptivity to congestion pricing among motorists confronted with congestion. What people say in answer to abstract questions in polls and how they react to actual proposals may well vary, but the poll data do imply an openness to consideration of the policy. Experience abroad suggests that gaining acceptance will not be easy; dedicating the revenues to purposes the public supports appears to minimize opposition.

Distributional Issues

As discussed in the previous chapter, all income groups can be compensated from the revenues gained by congestion pricing, but there would still be some adversely affected individual commuters. Although commercial activity on the whole would appear to gain from a more efficient transportation system, some businesses could be hurt. An economic analysis can demonstrate that congestion pricing would make society as a whole better off, but congestion pricing must be approved in a political context where the effects of policies on subgroups have considerable importance. Fundamental to congestion pricing's political feasibility is whether those adversely affected, even if a relatively small minority, would be more motivated and able to garner political support than those who would gain from the policy.

A review of the potential winners and losers helps illustrate the issues affecting political feasibility (Gomez-Ibanez 1992). In the case of tolling an existing road or bridge, the potential winners would be[1]

- Motorists with high values of time (the value of their time savings would exceed the cost of the toll),
- Travelers already using bus or high-occupancy-vehicle services (such users would benefit from time savings and the toll would be divided among several users),
- Most commercial users of the road system, and
- Recipients of the toll revenues (as described below, this could include road users, transit users, or taxpayers generally).

[1] In this list, Gomez-Ibanez's categories are expanded to include commercial users of the road system.

Among the potential losers would be

- Motorists with low values of time who stay on the tolled route for lack of alternatives,
- Motorists with low values of time who shift to an alternative, but less convenient facility,
- Motorists who forgo certain trips to avoid the toll, and
- Some commercial users and businesses.

In the political context of the debate on congestion pricing, the impact on those who would be "tolled off" among whom are low-income motorists, could have considerable influence on public opinion and political feasibility. The equity of shifting to congestion pricing is complex because the net impact on the poor depends on the equity of the existing system of transportation finance and consumption and how it would be changed by congestion pricing. It can be argued that the existing system is not very equitable, particularly in regions where regressive sales and property taxes are important sources of financing transportation facilities (and where middle-income drivers are the main users of new facilities being supported with sales taxes). Congestion pricing, in contrast, raises revenues directly from those users willing to pay for the service.

Determining the equity of the current system and how it would be changed by a shift to congestion pricing requires a detailed assessment of all sources of revenues and all benefits provided, with both contribution of revenues and consumption of services disaggregated by income group. Such an analysis is made difficult by the lack of data on important dimensions of these issues. The Environmental Defense Fund has conducted an analysis that was released after this report entered the National Research Council's peer review process (Cameron 1994). (The committee was briefed on the analysis but did not have an opportunity to review it in detail before its own report began the process of review, revision, and publication.) The committee did not take a position on the equity of congestion pricing. Rather, it limited its analysis to illustrating the change in economic impact across income groups. As outlined in the following sections, the distributional impact of congestion pricing would vary considerably depending on the design of the pricing system and the uses of the revenues to compensate those disadvantaged.

Areawide Pricing

Congestion pricing could generate substantial revenues. Regionwide congestion pricing for the greater Los Angeles area, for example, could generate over $3 billion annually (Small 1992). The decision about the use of the revenues earned from congestion pricing is critical to achieving the efficiency gains that motivate this policy. Congestion prices alone would dampen demand, which would relieve congestion, but unless the revenues were allocated well, average commuters would be somewhat worse off than before (Small 1983; Hau 1992). As illustrated in the preceding chapter, the fees required to reduce congestion would be larger than the time savings are estimated to be worth to middle- and lower-income motorists. Their time savings would be less than the out-of-pocket cost of the toll. If the revenues are well spent, the average commuter and society as a whole will gain.

The choice about how the money is spent depends in part on the goal being sought. If the goal is only to relieve congestion in both the near and the long term, the congestion fee should be set at a level that would be sufficient to provide for expected demand over the long term and the funds can be dedicated to providing the needed additional capacity (Keeler and Small 1977). Dedicating the revenues to maintenance and operation or added capacity, or both, would help convince motorists that the congestion fee is not just a tax that would be used for other purposes.

If the primary goal of using the revenues is to alleviate the distributional impacts, the distribution of the revenues could be quite different. One approach would be to use the revenues to offset other regressive taxes dedicated to transportation. For example, the reduction in sales, gasoline, and property taxes could be designed to benefit lower-income groups proportionately more than they would higher-income groups, thereby offsetting the disproportionate impact of congestion pricing on the poor.[2] If the disadvantaged groups include residents of a particular area, some part of the revenues could be dedicated to those residents in the form of tax rebates, commuter allowances, or improvements in roads, transit, bikeways, and sidewalks. If retail or commercial interests in specific areas would be negatively affected, specific funds might be targeted to improved transportation services for this group.

[2] The financing of transportation, and the burden borne by different income groups, varies widely across the nation. Property taxes are not necessarily regressive; many jurisdictions have some form of property tax credit that reduces tax rates for lower-income individuals. In addition, sales taxes, which do tend to be regressive, are not used to fund transportation in all metropolitan areas. In order to reduce the impact on the lower-income group in regions without regressive taxes, it might be possible to offer cash rebates that would diminish with family income or tax credits on a sliding scale, as is done with federal and state tax credits for child care.

Whether some or all of the funds should be dedicated to supporting transit through increased operating subsidies or capital grants opens up a complex debate. Many transportation economists view existing subsidies for transit with considerable skepticism (Lave 1991; Kain, Vol. 2). Although use of transit services for commuting has increased in cities such as New York and Washington, it has decreased in most metropolitan areas (DOT 1993). Suburbanization of jobs and residences combined with the increase in suburb-to-suburb commuting trips have decreased the share of the total commute market for which existing bus and rail transit provide competitive services (Wachs 1989). In most areas, congestion pricing could engender mode shifts to buses more than to trains. Bus transit could gain substantial new riders and revenues, thereby reducing the need for subsidy (Kain, Vol. 2). In other areas, however, congestion pricing could divert traffic from major arterials to minor arterials if these routes are unpriced. Because in these areas bus transit is more pervasive on minor arterials, an increase in congestion on these routes could degrade transit service.

The use of the revenues will no doubt be determined by the political bargaining required to help gain acceptance of congestion pricing; this bargaining will, in all likelihood, require a mixture of uses. For regional congestion pricing, it is quite possible to construct a program that would achieve the efficiency gains of congestion pricing and ensure that the majority of those disadvantaged by the fees receive equal or greater benefits in return (Small 1992). Such a hypothetical program, outlined in Table 4-1, has been designed for the greater Los Angeles area (assuming peak-period pricing on all thoroughfares). Under this program, two-thirds of the revenue would be returned to travelers and taxpayers in direct monetary payments (commuter allowances and property tax rebates) or tax reductions (fuel and sales taxes). The rest would be used to improve mass transit and build infrastructure (Small 1992). This program divides the revenues into three roughly equal parts. One-third would be returned directly to travelers. One-third would be returned to taxpayers generally through a reduction in the sales taxes and rebates on property taxes. The remaining one-third would be used for transportation, with equal parts being dedicated to highway and transit improvements and to transportation improvements in business centers that might be negatively affected by the congestion fee.

As shown in Table 4-2, this hypothetical revenue distribution would benefit commuters in lower-, middle-, and higher-income groups who

TABLE 4-1 Revenue Uses from Congestion Pricing Applied in the Los Angeles Region (Small 1992)

Program	Annual Amount ($ millions)
Reimbursements to travelers	
Employee commuting allowance ($10/month)	696
Fuel tax reduction ($0.05/gal)	348
General tax reductions	
Sales tax reductions (half of transport surcharge)	525
Property tax rebate (eliminate local highway subsidy)	465
New transportation services	
Highway improvements	315
Transit improvements	312
Transportation services to business centers	322
Total net revenue	2,983
Collection costs	137
Total gross revenue	3,120

have commutes of average length [16 km (10 mi)]. Solo drivers in all three groups come out slightly ahead. Those who switch to transit or carpools would benefit even more. Lower-income solo driver commuters who switch to transit and those who were transit riders before and after congestion pricing show the largest gains as a percentage of income.

Not everyone would benefit from this example, however. If an average toll of $0.09/km ($0.15/mi) applies for the entire length of a freeway trip, commuters with longer-distance trips will pay considerably more than those with a commute of average length. Roughly one-fourth of commuters in Southern California have a one-way commute of 40 km (25 mi) or more (Giuliano, Vol. 2). Middle-income and lower-income solo drivers with trips of 40 km would be net losers in the foregoing pricing and revenue distribution system (Giuliano, Vol. 2, Tables 7 and 8). However, both middle- and lower-income solo drivers who switch to carpools or transit would become net winners. Although solo drivers making long-distance commute trips might represent a minority of commuters, they might well resist shifting mode and might be vocal in opposition.

Facility Pricing

As in the case of areawide pricing, congestion pricing on individual facilities should result in net benefits to commuters and an efficiency gain for the local economy. Facility pricing raises issues of political feasibility that

TABLE 4-2 Impact of $0.15/mi Congestion Fee in Southern California for Commuters with Average 16-km (10-mi) Trip

	Annual Net Benefit[a]					
	Middle Income[b]		Higher Income[b]		Lower Income[b]	
Commuter Status	Dollars	Percent of Income	Dollars	Percent of Income	Dollars	Percent of Income
Solo driver before and after imposition of toll	166	0.7	546	1.4	55	0.4
Solo driver to two-person carpool	528	2.2	644	1.6	470	3.1
Solo driver to bus transit	683	2.9	754	1.9	907	5.9
Carpool before and after imposition of toll	531	2.3	784	2.0	397	2.6
Bus transit before and after imposition of toll	906	3.8	1,159	2.9	772	5.0

SOURCE: Based on data from Giuliano (Vol. 2, Tables 7 and 8) and Small (1992), Table 2.

[a] After paying toll, valuing time savings, and redistributing revenues. Revenue distribution based on Table 4-1 and data from Giuliano (Vol. 2, Tables 7 and 8).

[b] Income of wage earners selected to represent annual wages of average, 80th-percentile, and 20th-percentile workers based on hourly wages of workers in the Los Angeles region (Small 1992, footnotes 17 and 18). In annual terms these wages would be roughly $24,600, $40,100, and $15,300.

are less central in the context of areawide pricing. Pricing on individual facilities, for example, is more likely to have direct impacts on travelers originating from or destined to specific geographic areas. When tolls affect some individual areas or jurisdictions more than others, the local politics of compensating the more affected groups becomes more complex. The issues of raising tolls on tolled facilities serving New York City or placing tolls on some of the free bridges serving Manhattan have become enmeshed in regional political struggles (Zupan, Vol. 2). Commuters originating from specific neighborhoods that use a facility to be priced would be faced with paying the fee, using an alternate route, or making some other arrangements. The shift in traffic from priced to unpriced facilities (and potential increases in congestion on unpriced routes) and the possible diversion of traffic to residential streets would also affect political feasibility.

The potential revenues from congestion pricing could be used to provide alternatives for those affected. Preliminary estimates of the revenue gain from a $1.00 toll increase for the Bay Bridge, for example, suggest increased annual revenues of about $8 million, or a 14 percent increase in total toll revenues. Redistributing the funds broadly according to the hypothetical example outlined in Table 4-1, however, would not effectively target those most affected. In the proposal for the Bay Bridge, the funds will be dedicated to improving alternative modes of transportation in the same corridor. Planners are also considering an off-peak discount to provide automobile commuters with the alternative of shifting the timing of their trips.

Parking Pricing

The potential winners and losers from a regional parking pricing strategy have not been examined as extensively as the winners and losers in the above examples. Such a proposal would probably have distributional consequences similar to those of an areawide pricing strategy for roads. The time savings for commuters, however, would be smaller because parking pricing would not affect through travel (see Tables 3-1 and 3-2).

A variation on parking pricing, a "cash-out" of employer-paid parking, has been proposed that ensures that almost everyone benefits. California has enacted such legislation to require employers who lease parking spaces for their employees to provide commuters with the choice of taking the employer-subsidized parking benefit in cash (Shoup, Vol. 2).

The California "cash-out" law would have a limited impact. It applies only to existing leases for parking. Many, if not most, leases combine parking and office space. In order to affect a larger share of parking, the law could be expanded to require property owners to separate office space and parking when leases are renegotiated.

The cash-out option is proposed in an attempt to reduce the incentives for solo driving that are incorporated in the federal tax code, which treats employer-provided parking as a tax-exempt fringe benefit. Although the cash-out option is fairly simple in concept, it becomes more complex in implementation. It would be more expedient to simply eliminate the tax-free benefit of employer-provided parking, but this would be difficult to achieve politically. The cash-out option is designed specifically such that it provides net social benefits while taking little, if anything, from anyone, which enhances its political feasibility (Shoup, Vol. 2). Taxing employer-provided parking would probably have a much larger effect on solo driving, but would be more difficult politically because it would be taking away a tax benefit enjoyed by millions of workers.

Revenue Use

Not all the diverse uses of the funds discussed above as ways of ensuring that all income groups benefit would be allowed in the current congestion pricing pilot program authorized under the Intermodal Surface Transportation Efficiency Act of 1991 (ISTEA). The legislation limits uses of the funds raised in pilot projects to purposes eligible under Title 23 of the federal code. This effectively limits the funds to specific transportation purposes permitted under federal law (which has been interpreted by the U.S. Department of Transportation to include transit operating funds). This legislation appears to be unduly restrictive. Whereas the seed money made available by the federal government might be appropriately restricted to purposes eligible under Title 23, the funds that are collected in the form of congestion tolls are local funds, and local governments ought to be able to apply them where they are deemed necessary. Indeed, more flexibility may be required in order to give local governments sufficient latitude to form coalitions that support congestion pricing.

One can imagine an even broader use of funds than those already discussed. For example, a local government could use some portion of the revenues earned from congestion pricing to help rebuild distressed areas or subsidize the day-care costs of low-income working parents. The federal restrictions were apparently imposed in part to keep highway funds from becoming "cash cows" for local government. There are, however, very practical constraints on how congestion pricing revenues could be used. The voters, almost all of whom are road users, will ultimately decide whether any form of tolling will be put in place, and the feasibility of gaining public approval will be highly dependent upon the uses of the revenues. Public acceptance is most likely to be gained by proposals that would use funds in ways that benefit highway users directly, for example, by maintaining or providing capacity. The public might also accept revenue uses less directly related to their own transportation needs but still applicable to concerns about equity, such as subsidizing special transportation services for specific geographic areas. To the extent that proposals for revenue uses appear to highway users and voters to be supporting general government purposes (such as schools and other social services), there is the risk that the public will perceive the congestion toll as another form of taxation. Of course, if the motoring public approves supporting these purposes from congestion fees, should they not be allowed to pay for them with locally collected revenues?

Summary

It is possible to use the revenues earned from congestion pricing to compensate the losing groups such that each group benefits. The question of political feasibility depends in large part on how intensely individuals would experience the gains and losses. For example, motorists with high values of time, bus and high-occupancy-vehicle users, and recipients of the revenues may be winners, but since these gains are spread among many individuals, many of whom may receive relatively small benefits, these winners may not be sufficiently motivated to fight for the benefits. Disadvantaged individuals, even assuming some compensation, may still feel sufficiently aggrieved to resist any change forcefully. Politicians are reluctant to espouse policies that will have widely dispersed benefits and highly concentrated costs (Rom, Vol. 2).

Although most middle-income users would benefit, some would not, and some could be expected to oppose congestion pricing as a general concept. Giuliano (1992) suggests that middle-income users opposed to paying congestion tolls would use the possible impact on the poor as an argument against the impact on themselves. In a political context, the equity issue could be used by opponents as a wedge to fragment coalitions that might otherwise support congestion pricing.

Building Support

As with individual opposition, groups opposed to congestion pricing are likely to be more vocal than those that support the concept (Rom, Vol. 2).

Political feasibility depends on the influence of the opposing groups on the public and elected officials. Political leaders could pursue at least three generic strategies to address the influence of opposed groups (Rom, Vol. 2): (*a*) narrow strategies, which would design programs to minimize the negative effects on specific groups; (*b*) broad strategies, which would build support by accentuating the positive consequences of congestion pricing among groups not opposed to congestion pricing to offset the opposition by groups resisting change; and (*c*) transformational strategies, which would seek common ground among groups. Political leaders could pursue these strategies singly or in combination.

Pursuit of a narrow strategy is implicit in the revenue distribution proposals discussed above, in which negatively affected groups are directly compensated. Design measures could also be used to reduce negative effects. For example, resistance to congestion pricing by neighborhood groups concerned about diverted traffic could be minimized by prohibiting through traffic from using residential streets during peak periods or by altering road designs (narrowing lanes) to make passage by through traffic more difficult.

A broad strategy would build coalitions among those interest groups that would support some of the benefits of congestion pricing. In California, for example, large employers supporting congestion pricing as an alternative to employee trip reduction programs, commercial developers supporting congestion pricing as an alternative to impact fees, and environmental groups supporting policies to reduce automobile dependence have formed an influential constellation of interests in support of congestion pricing (Giuliano 1992).

Transformational strategies would be more difficult in the case of facility or areawide congestion pricing because, despite the net benefits to society, not everyone wins. Cashing out employer-provided parking subsidies is an example of a transformational strategy, because most parties win. Another possibility deserving further research is to have a system of tradable permits for peak-period driving (Rom, Vol. 2). In this approach, all residents would be given permits to travel, which they could then sell to the highest bidders. This would allow motorists to be "bought off" the roads during peak period rather than being "tolled off." This approach might even be further modified to make it more progressive by providing marketable permits, the nominal value of which would be inversely proportional to family income. Although the idea of tradable permits is intriguing, it raises many questions about how such a system would be designed and implemented. The committee has not attempted to evaluate this proposal. Research is needed.

Public education can also play a role in building support by communicating important ideas and values (Rom, Vol. 2). Rather than attempting to restrict choices through regulation, congestion pricing would provide net benefits by increasing economic efficiency and productivity, saving time, and providing motorists with choices about when to drive. Public support is also likely to grow if incremental steps toward congestion pricing prove effective and the experience with early experiments is disseminated widely.

The promotion by political leaders of any of the strategies identified above depends upon first gaining their interest and support. Reforms such as congestion pricing require strong constituencies devoted to fighting for change at each of the many points in the American political system where opposing groups can exercise veto power. Without strong constituencies, or without the belief that such constituencies could be developed, political leaders are unlikely to promote congestion pricing on their own. The alliance between an interest group and a political leader is generally a prerequisite to embarking on a campaign for change.

Giuliano (1992) has identified three interest groups that have coalesced in California in favor of congestion pricing: large employers opposed to mandates that they manage their employees' commuting behavior, commercial developers, and environmentalists. To this list could be added high-value-of-time highway users and the potential recipients of the revenues. For each of these groups there are reasons to support congestion pricing. Large employers could avoid the cost of employing travel demand managers. Commercial developers could avoid impact fees on new development (which require the developer to contribute to the cost of improving highway facilities). Environmentalists could be attracted to the potentially large changes in emissions that congestion pricing offers relative to other demand management measures. High-value users currently forced to operate under congested conditions could save time and money. Revenue recipients might include transit agencies that are struggling to find a replacement for federal operating subsidies (which are declining in real, if not absolute, terms), those with low incomes who do not drive much during peak hours, state departments of transportation, local governments, and groups opposed to higher property and sales taxes.

Whether these groups perceive that they could gain sufficiently from congestion pricing such that they would vigorously support it is a key question in determining political feasibility. Certainly in the San Francisco

Bay Area, business leaders have strongly advocated congestion pricing through the Bay Area Council, and it is possible that similar support will emerge in the large metropolitan areas required by the Clean Air Act Amendments of 1990 to develop employer trip reduction plans. Similarly, the interest of commercial developers will be conditioned by the presence of less palatable alternatives, such as the imposition of impact fees to provide for new capacity. Although environmentalists have certainly helped revive interest in congestion pricing, the issue could not be said to have widespread support as a policy instrument among the different environmental groups. Some environmental groups find regulatory approaches to dealing with pollution more appealing than marketplace approaches. Some may remember the backlash that occurred in the 1970s when environmental groups supported transportation pricing measures to meet the requirements of the 1970 Clean Air Act. As for high-value-of-time highway users, the question is partly one of whether they have a local organization that would represent their interest on this issue. The local clubs of the American Automobile Association, for example, have long opposed tolling of any kind, as have the various state members of the American Trucking Associations. In contrast, the potential beneficiaries of the revenues, which might include transportation agencies, are well organized and might well be willing to lobby for congestion pricing. Transit agencies in particular might benefit from additional ridership and by local efforts to ensure adequate alternatives for motorists "tolled off" the highway.

The support of constituency groups willing to be advocates for congestion pricing is important because the American political system is organized in a way that makes change difficult. Proponents of change must be willing to advocate their cause successfully before state agencies, city and county councils, state legislative committees, and high-level government officials. Opponents of change, however, can stymie, if not block, progress by winning at any of these points.

SUMMARY

Advances in technology are rapidly overcoming the perception of problems with congestion pricing associated with queues of motorists at toll booths. Innovations in technology are also providing options that would lessen concerns about the invasion of privacy.

Although there does not appear to be an ideal institutional mechanism for managing a regional congestion pricing program, workable institutional forms exist or could be adapted to this purpose. This does not mean that adoption of regional congestion pricing would be a simple matter. For example, a large-scale distribution of revenues could be required in a regional congestion pricing program. The distribution could also be quite complex because of the cross-jurisdictional flows of traffic and the potential need to compensate motorists who would originate in one jurisdiction but who would be charged a fee in another. The public would need to be convinced that a (potentially new) government agency could successfully and fairly manage a change of this magnitude. Because of the magnitude of the change and the complexities involved in a regional program, the feasibility of pricing individual facilities within the next few years is far greater than that of a proposal for regional pricing. For pricing individual facilities, these institutional forms are likely to vary widely across regions depending on existing local institutions, practices, and laws.

Regarding political feasibility, the average U.S. motorist is clearly not prepared for a radically different method of charging for road use, but opinion polls suggest that motorists already frustrated with congestion would be receptive. In parts of California, where funds are scarce, needs to address air quality problems are pressing, and the alternatives to congestion pricing are unappealing, interest is growing. When congestion is severe and the alternatives are unpalatable, certain forms of congestion pricing have become more acceptable.

The perception that congestion pricing is unfair to specific groups, particularly those unable to pay without economic hardship, is probably the most significant problem in its political acceptability. Programs can be designed to compensate all disadvantaged groups through specific policies for allocating the substantial revenues that can be earned. Inevitably, some individuals will be disadvantaged. Some of the concern expressed about the impact on the poor, however, may be less a genuine concern about equity than a political strategy by groups opposed to paying congestion fees.

Political leaders can pursue different strategies to enhance the feasibility of congestion pricing by minimizing opposition through program design, by maximizing support among interest groups that would support congestion pricing for reasons other than congestion relief, and by communicating essential ideas and values that congestion pricing proposals encompass.

Given the nascent state of congestion pricing as a policy instrument, it would be premature to expect local political leaders to champion such a policy. However, in areas with rising congestion, concerns about air quality, constraints on capacity expansions, and a series of unattractive options for dealing with these problems, congestion pricing could gain greater interest and appeal, particularly if the early trials with it are successful.

REFERENCES

ABBREVIATION

DOT U.S. Department of Transportation

Apogee Research, Inc. 1990. *The Road Information Program (TRIP) National Transportation Survey: 1990 Poll Results*. Summary Report. Road Information Program, Washington, D.C.

Bay Area Council, Inc. and KQED, Inc. 1991. *Bay Area Poll*. San Francisco, Calif.

Borins, S. 1988. Electronic Road Pricing: An Idea Whose Time May Never Come. *Transportation Research*, Vol. 22A, No. 1, Jan., pp. 37–44.

Cameron, M. 1994. *Efficiency and Fairness on the Road: Strategies for Unsnarling Traffic in Southern California*. Environmental Defense Fund, New York City, Feb.

DOT. 1993. *Journey-to-Work Trends in the United States and Its Major Metropolitan Areas, 1960–1990*. Final Report.

Downs, A. 1992. *Stuck in Traffic: Coping with Peak-Hour Traffic Congestion*. The Brookings Institution, Washington, D.C.

Giuliano, G. 1992. An Assessment of the Political Acceptability of Congestion Pricing. *Transportation*, Vol. 19, No. 4, pp. 335–358.

Gomez-Ibanez, J. 1992. The Political Economy of Highway Tolls and Congestion Pricing. *Transportation Quarterly*, Vol. 46, No. 3, pp. 343–360. (For summary, see Presentation Summary, In *Exploring the Role of Pricing as a Congestion Management Tool*, Searching for Solutions: A Policy Discussion Series, No. 1, U.S. Department of Transportation, March.)

Gomez-Ibanez, J., and K. Small. Forthcoming. *NCHRP Synthesis of Highway Practice: Road Pricing for Congestion Management: A Survey of International Practice*. TRB, National Research Council, Washington, D.C.

Gomez-Ibanez, J., J. Meyer, and D. Luberoff. 1991. The Prospects for Privatizing Infrastructure. *Journal of Transport Economics and Policy*, Sept., pp. 259–278.

Hau, T. 1992. *An Economic Analysis of Road Pricing: A Diagrammatic Approach*. Policy Research Working Paper. WPS 1070. The World Bank, Washington, D.C.

Jones, P. 1991. Gaining Public Support for Road Pricing Through a Package Approach. *Traffic Engineering and Control*, April.

Keeler, T., and K. Small. 1977. Optimal Peak-Load Pricing, Investment, and Service Levels on Urban Expressways. *Journal of Political Economy*, Vol. 85, No. 1, pp. 1–25.

Lave, C. 1991. Measuring the Decline in Transit Productivity in the U.S. *Transportation Planning and Technology*, Vol. 15, pp. 115–124.

Roth, G. 1991. Pricing, Financing and Ownership of Roads in a Free Society. Presented at Second International Conference on Privatization and Deregulation in Passenger Transportation, Tampere, Finland, June 16–20.

Small, K. 1983. The Incidence of Congestion Tolls on Urban Highways. *Journal of Urban Economics*, Vol. 100, No. 4, pp. 888–898.

Small, K. 1992. Using the Revenues from Congestion Pricing. *Transportation*, Vol. 19, No. 4, pp. 359–382.

Wachs, M. 1989. U.S. Transit Subsidy Policy: In Need of Reform. *Science*, Vol. 244, pp. 1545–1549.

5

Design, Evaluation, and Research

Although relatively few proposals for congestion pricing are moving forward in the United States, there appears to be growing interest. The proposals that are enacted will provide unique opportunities to learn about what will and will not work. A poorly designed initiative that fails, however, may spell the end of this approach to managing congestion or at least set it back by another 10 to 20 years.

Offered in the first section are a few design issues that need to be considered in any congestion pricing proposal. Given the lack of experience with congestion pricing in the United States, there are many questions about design that do not have clear answers. Learning from early efforts is critical to determining the effectiveness of this approach and for improving future proposals.

In the second section, the basic features that should be part of the evaluation of these early experiments are outlined, and in the final section, other important research is suggested that would improve the understanding of congestion pricing.

DESIGN ISSUES

Several issues have been discussed in previous chapters that are critical to the design of any congestion pricing project, such as mitigating distributional consequences and the use of the revenues earned. Discussed in this section are several other issues that need careful consideration: allowing sufficient time for a proposal to develop political acceptance, setting an appropriate price, and providing alternative routes or modes of transport for those who prefer not to pay a congestion fee.

Proposal Development

Congestion pricing has gained currency only in a few places, and even in these cases the process of gaining political acceptance has been quite slow and conditioned by almost unique circumstances. Interest in California has surely been fostered by the fiscal restrictions that voters have placed on state and local government spending, the continued rapid growth in population and automobile travel, and the stringent requirements of the California Clean Air Act. In the San Francisco Bay Area, congestion pricing has also been furthered by broad public concern about the environment and by the sophisticated advocacy of a business-sponsored organization dedicated to resolving regional issues in the Bay Area (Dittmar, Vol. 2). Even with these conditions influencing adoption of congestion pricing, the Bay Area congestion pricing proposal was some years in the making (and at the time of this writing requires state enabling legislation to raise the toll). In San Diego, interest in congestion pricing was fueled by accelerated growth in traffic congestion over the last few years, the leadership of local elected officials, and the general acceptance of a philosophy of charging the user directly for the cost he or she imposes on others (Duve, Vol. 2). Even so, acceptance of the concept of congestion pricing emerged over a considerable period of time.

In contrast, an effort to build consensus for a congestion pricing proposal for Manhattan during the late 1980s was summarily dismissed by the public and by some downtown commercial interests (Altshuler 1990; Zupan, Vol. 2). These cases suggest that interest in congestion pricing—driven by concerns about congestion and the environment—will emerge in a few places, but sufficient public and political acceptance to allow it to go forward will occur slowly, if at all. The congestion pricing pilot projects allowed under the Intermodal Surface Transportation Efficiency Act of 1991 (ISTEA) may be developed at any time during the 6-year authorization period, but the time frames for development and implementation of proposals to meet the first and second rounds of solicitations assumed an ability to move forward at a speed that few jurisdictions could meet.

In the interest of political feasibility, the development of congestion pricing proposals requires time to design features that address the concerns of groups resisting change. In both San Francisco and London, detailed studies of the possible impacts and costs of congestion pricing are being used to further public awareness and acceptance. In addition, the ongoing London study (a 3-year, $5 million effort) includes a considerable effort to gauge and inform the opinions of local leaders as well as the public at large (Gomez-Ibanez and Small forthcoming). Such efforts highlight the difficulty of convincing political leaders and the public of the potential of congestion pricing and the considerable effort required to move proposals forward. Because of the need to build consensus slowly and to fashion a politically feasible congestion fee, the first year of the San Francisco Bay Area congestion pricing pilot project will be devoted to extensive analyses of travel patterns and how they might be affected by toll increases (with particular emphasis on the impact on low-income individuals), consideration of off-peak discounts for freight operations, estimates of the potential mode shifts to transit, focus group and survey research of potential public reaction to different price increases, and review of the proposal with various interests that could benefit from or block the proposal.

Setting the Price

Setting a proper price is important to the success of congestion pricing. If the fees are too low, they will not divert sufficient traffic to relieve congestion, which would undermine the efficacy of this proposal in the minds of motorists and voters. If the prices are too high, the road system will be underused, and motorists and voters can be expected to be critical of that as well. One of the criticisms of Singapore's area licensing proposal is that it set the fees too high initially (Toh 1977). Whereas this greatly reduced automobile congestion in the downtown area, it apparently diverted too many travelers to facilities unable to accommodate them, such that little or no improvement occurred in total trip times to the core (Gomez-Ibanez and Small forthcoming).

Given the lack of experience with congestion pricing in the United States, however, there is no perfect way to predict the appropriate prices to charge (Vickrey 1993). Vickrey (1993) and Wohl and Hendrickson (1984) suggest deriving an estimate for the peak fee using standard approaches to calculating demand elasticities and short run marginal costs, and then subsequently adjusting the fees in response to observed traffic patterns. With the administrative and political constraints that congestion pricing proposals face, however, it may be difficult to obtain public and political approval of a proposal without publishing the fees in advance.

One way to predict the right price is first to determine an appropriate traffic flow for a facility and then to set the price according to estimates of motorists' sensitivity to price increases in analogous situations. Although the models applied to develop this estimate are quite complex, this is essentially the method used in estimating the benefits of congestion pricing for the San Francisco Bay Area and Southern California (Harvey, Vol. 2). Another method for setting the right price is to estimate those costs and benefits that can be quantified and then to calculate the price level that maximizes the net benefits. This approach is being used in the London study currently under way (May, Vol. 2).

In the case of the Riverside Freeway project in Southern California, the private developers are being allowed to set and vary the prices according to demand, with no prior commitment to the public about what those prices might be. Rather than attempting to regulate the prices that the developers could charge motorists, the state chose to regulate the maximum return that developers could receive on their investment (Fielding, Vol. 2). The developers will obviously have to be sensitive to consumer demand in setting and varying their prices and are also required to maintain free-flow conditions for high-occupancy vehicles. Because this is a private project, however, there is less public information available about how the developers will estimate demand and how frequently the prices might be changed.

During the first year of the San Francisco Bay Area congestion pricing pilot project, public and political opinion will be surveyed on the receptivity to different price increases for the Bay Bridge tolls. It appears that the price increase for the Bay Bridge will be on the basis of both extensive modeling of the price increase required to improve traffic flow to a desired level of service and an appreciation of what is politically feasible.

Another pricing issue to be confronted is whether the structure of the congestion fee would be a simple, flat increase or smoothly varying increases. For the latter, a congestion fee might increase 10 percent at the beginning of a morning peak hour and then increase substantially in 15-min increments to its highest level when traffic reaches its normal peak. The toll might then decrease gradually during the shoulder of the peak period. The developers of the Riverside Freeway project intend to use a smoothly varying price structure.

Flat congestion tolls lasting for the duration of the peak period have the advantage of being simple to understand, but have other disadvantages. Motorists who typically travel during the middle of the peak but who would prefer to shift to an earlier time to avoid the congestion fee may have to adjust their departure by as much as an hour. A flat fee structure could incorporate a substantial off-peak discount to induce such adjustments, but such a large time shift may reduce motorists' willingness to change.

With a smoothly varying toll, in contrast, motorists could adjust their departure time in order to pay the level of toll they find acceptable. In theory, if motorists all make small adjustments according to their differing values of time, substantial time savings would occur for all without requiring a few motorists to make very large time adjustments (Vickrey 1993). Another disadvantage with a flat toll is that some motorists who want to avoid the toll will arrive just before or just after the change and will thereby increase congestion. In Singapore, for example, during the half hour before the congestion fee took effect, congestion initially increased by 24 percent (Watson and Holland 1978, 43).[1]

A major question about variable tolls is whether drivers will be able to predict with sufficient certainty the fee that they will face when they reach the tolled facility, thus ensuring the traffic smoothing predicted by theory. There is no technology on the horizon that would predict, in real time, the congestion that a traveler might encounter during the trip and the fees that the traveler might pay (May, Vol. 2). The actual effects on motorist behavior and traffic flow in response to the variable pricing proposed for the Riverside Freeway (SR 91) will be of great interest in this regard.

A number of different variable pricing proposals have been made in an attempt to optimize the efficiency gains of congestion pricing. Additional proposals to the ones outlined here, and the possible technologies for implementing them, are reviewed in the paper by May in Volume 2 of this report.

[1] Because the increase represented a relatively small number of vehicles, the authorities did not believe that the increased traffic represented a substantial problem. A similar increase in traffic initially occurred just after the peak-period fee ended. In this case the authorities extended the period during which the charge was imposed until mid-morning.

Available Alternatives

Travelers will make adjustments to congestion pricing in ways that cannot be fully predicted. Assuredly, however, some will seek alternatives to avoid the fee. Underestimating the size of these shifts could result in substantial dislocations. In Singapore, for example, it appears that the high price for entering the central area diverted a substantial portion of through traffic to a ring road that could not handle the capacity. Planners also underestimated the shift to transit. The large number of new transit patrons apparently increased bus boarding times and therefore slowed the travel times of the buses (Gomez-Ibanez and Small forthcoming). On the other hand, planners overestimated the demand for park-and-ride facilities on the edge of the central business district, which resulted in construction of some expensive and unnecessary parking facilities and purchase of small buses to operate between the lots and the city center. In the aggregate, the congestion on the ring road and slower travel times for bus patrons negated many of the time savings enjoyed by those who paid the fee to travel into downtown. Over time, the net benefits became more positive with the expansion of the ring road and bus fleet (Gomez-Ibanez and Small forthcoming), but the dislocations caused early in the Singapore experience suggest that considerable care should be paid to providing alternative modes or routes. In the example of the A-1 motorway in France, considerable care was taken to minimize any tendency to divert travelers to adjacent, signalized arterial routes.

When congestion pricing is applied to a single facility, there can also be the problem of traffic diverting from a tolled route to untolled routes. This diversion may slow traffic on the untolled routes. Motorists may also begin using neighborhood streets to avoid tolled routes, and fear of such effects by neighborhood groups could lead to rejection of proposals for congestion pricing. Any proposal for congestion pricing should make explicit the plans for handling traffic diversion and incorporate design features to minimize its impact. Avoiding the creation of congestion on unpriced routes would be particularly important for avoiding localized increases in automotive emissions.

EVALUATION

Several key policy questions emerge from the previous chapters of this report, the answers to which will only be gained through experience. The major questions about congestion pricing can be summarized as follows:

1. What is the range of behavioral responses at different prices?
2. How will behavioral changes affect congestion?
3. What will the impacts be on different groups (considered by income, gender, and geographic area)?
4. What will the air quality and energy effects be?
5. What will the effects on urban form and development be?
6. How do all of the above affect public receptivity and political feasibility?

In the sections that follow, some basic evaluation elements are outlined that could be employed to learn substantially more than is now known about each of these questions. In an individual corridor, in which a priced route would be parallel to a free road, an evaluation could gauge fairly readily motorists' willingness to pay to avoid congestion. The privately financed and operated toll road in the median of the Riverside Freeway, which will have congestion tolls for single-vehicle occupants, offers just such an opportunity. The extensive evaluation of the Singapore congestion pricing program, conducted by staff at the World Bank with financial support from international organizations and from the Environmental Protection Agency, U.S. Department of Transportation, and U.S. Department of Energy, provides a useful framework (Watson and Holland 1978); the following text draws upon that design.[2]

Travel Behavior

Questions about how personal, business, and commute trips would be affected by congestion pricing are among the most intensely debated within the transportation community, and for which, aside from mode shifts between automobile and transit, little empirical information is available (Kain, Vol. 2). Very little information is available at present to guide estimates of how shifts in the timing of trips would affect the total loading on the transportation system. This is a particularly important area for evaluation. One approach to measuring travel behavior is to collect detailed household travel diaries for some time period (24 hours or 7 days). This was the approach taken in the Singapore study and in the various personal transportation surveys undertaken by the Bureau of the Census.

---

[2] The Federal Highway Administration has contracted for a guidebook on project development and evaluation (Bhatt forthcoming).

Travel diary data are expensive to collect, however, which helps explain why such surveys are conducted infrequently. The Singapore study, for example, collected before-and-after travel diaries from a sample of 2,000 travelers (the "after" survey results were supplemented with less extensive surveys of another 10,000 residents). These data allowed evaluators to measure changes in overall trip making, mode shifts, and elapsed travel times during the peak, and route shifting. All these elements were important to gauging whether the program was a success or failure. Less extensive information about travel behavior can be collected through telephone surveys. Either approach would allow for an estimate of demand elasticities, trips forgone, trip chaining, shifts to other modes, and the use of travel substitutes, such as increased telephone use, catalogue ordering, and telecommuting.

The World Bank evaluation of Singapore relied on traffic counts of commercial vehicles, but such data are not likely to address the ways in which commercial activity might be affected positively and negatively. It would be much more useful to draw a sample of carriers and business entities dependent on the transportation system and collect travel and cost data before and after congestion pricing was put in place.

Effects on Facilities

Another of the most important questions is whether and how much congestion pricing reduces congestion on specific facilities. Traffic counts and vehicle speeds could be collected by time of day and day of week on tolled and untolled routes with the automated technologies routinely used for these purposes. Because of seasonal and economic effects on travel, however, it would be important to develop sufficient trend data to control for these external effects on travel behavior and to take repeated measurements of traffic effects over several months or quarters after a new policy was put in place. Simple before-and-after surveys could easily be distorted by seasonal effects unless they were carefully timed to occur at the same time of year, but even this degree of care would not be sufficient to control for other external effects on traffic. In addition, the before-and-after approach does not address what the traffic would have been in the absence of the intervention.

Automated traffic counts will also not provide measures of changes in vehicle occupancy and in the composition (personal/commercial) of vehicular traffic. Measures of such changes can be inferred from the travel diary data but should be confirmed by visual observation.

In cases where transit options are available to travelers, shifts to transit can be estimated from travel diaries but should be confirmed with ridership surveys. Observations of sample routes and trips would also provide important information about increased crowding, if any, and the effects this might have on speed.

The effects of congestion pricing on transit service would be of considerable interest. The improvement in traffic flows would provide faster and more timely bus service; this and the direct monetary incentive for people to shift from automobiles to transit should improve ridership and revenues (Kain, Vol. 2). Such improvements would benefit the lower-income riders who rely on this service and could mitigate any adverse effect congestion pricing might have on the poor. The effects on transit service reliability and frequency should be closely monitored.

Impacts on Specific Groups

A panel survey, in which data would be collected from travel diaries or telephone surveys, would enable researchers to address many of the questions about the equity consequences of congestion pricing. The data collected should include important household and life-style characteristics: income, residential location, commuting pattern, number of vehicles, number of workers, family size, children's ages, annual travel, and employment type and location. Without these data, it would be difficult to measure the consequences on groups that could be most affected.

Environmental and Energy Consequences

Given the concerns about the environment and energy conservation that help motivate the interest in congestion pricing, it is important to collect data on both travel behavior and actual emissions. For example, some estimates that congestion pricing would produce energy and environmental savings assume a net decline in travel, but if a substantial share of trips are diverted to other times, the environmental and energy benefits would be lower. Data collected from travel diaries or surveys would provide an assessment of how the pattern of travel activity was affected by congestion pricing.

Congestion pricing is assumed to improve the flow of traffic and thereby reduce emissions because of less stop-and-go traffic. Such effects can be inferred from traffic engineering surveys of vehicular flow, but the actual measures of emissions on the road during peak periods would also be important. When individual facilities are tolled, for example, traffic diverted to other routes can increase the congestion and emissions on those routes. Whether congestion pricing results in net environmental benefits requires an empirical answer. New technologies, such as the Stedman scanning device, allow emissions to be measured while vehicles are in use (Bishop et al. 1993). The Stedman device directs an infrared beam through the exhaust emission of individual vehicles (or a traffic stream) to a measuring device. The characteristics of the light indicate carbon monoxide and hydrocarbon content. This kind of technology would permit sampling of vehicle emissions before and after the toll was put in place. It is also important to measure changes in concentrations of pollutants, particularly carbon monoxide. Emission monitors could measure changes in emissions on tolled facilities before and after the tolls were put in place. Data from these monitors, combined with traffic counts and traffic flow estimates, would allow inferences to be drawn about the net effects on pollutants.

Effects on Urban Form

The effects that congestion pricing might have on urban form are among the most difficult to measure. Changes in business and personal location occur over several years, during which time the effect of congestion pricing could be tempered by a number of other countervailing influences. Although congestion pricing might lead to more dense development near activity centers, such changes take time and would often be affected by zoning and development restrictions. It would be useful to conduct surveys of managers of commercial activities to determine whether or how they might react, but it is difficult to know what managers will actually do compared with what they think or say they might do. Perhaps more useful would be research on how to properly evaluate the impact of congestion pricing on urban form.

Political and Public Receptivity

The perception of how a policy intervention works may be as important to its continued operation as the actual effects it has on congestion. Such opinion surveys could also provide insight into issues such as privacy concerns. Specific attention should be given to sample design to ensure that it captures those most affected (for good and ill), but also to ensure that the sample in the aggregate is representative of users of the facility.

Summary

Careful and extensive evaluation of early congestion pricing programs is critical to developing a better understanding of the advantages and disadvantages of this policy. An appropriate evaluation of a congestion pricing project must be designed carefully to account for different explanations of the possible outcomes other than the introduction of a congestion fee (Campbell and Stanley 1966; Weiss 1972). Considerable data must also be collected. The evaluation outlined in this chapter could easily extend over several years and cost $5 to $10 million. The costs would be high, but it would be a worthwhile public investment. Congestion wastes billions of dollars in metropolitan areas every year, and it has persisted despite the expenditure of many billions of dollars to expand highway and transit capacity. Many of the transportation control measures being considered and implemented to meet clean air standards are estimated to have quite modest effects on travel (Deakin 1993). Despite its political challenges, congestion pricing has more promise than most other policies for significantly reducing congestion. The expense of a few million dollars to demonstrate the potential of the theory would be a good social investment.

OTHER RESEARCH

Although many of the most important questions about congestion pricing—the time savings and potential adverse impact on the poor—can only be answered through evaluation of actual projects, there are issues and questions about congestion pricing that deserve further exploration independent of any project actually going forward.

Commercial Transportation and Activities

In examining the possible impact of congestion pricing on business activity and commercial transportation in the United States, the committee found no prior studies to inform its analysis. Given the importance of transportation to the economic vitality of metropolitan areas and the widespread perception that congestion increases business costs, it is remarkable how little is known about the magnitude of these costs. Surveys of service firms in congested metropolitan areas reveal that the managers of these firms believe that congestion affects them but have little specific information (Cambridge Systematics forthcoming). Transportation planners have little information about the logistics patterns of businesses dependent on transportation or the costs borne by commercial carriers operating in congested metropolitan areas (Cambridge Systematics forthcoming; Giuliano, Vol. 2). Research is needed on the logistics patterns of metropolitan firms; the frequency, origin, destination, and timing of trips; substitutes for transportation; and the ability of firms to adapt to congestion without affecting costs. All of these elements are important to predicting how firms would be affected by congestion pricing. Research is also needed on how congestion pricing would affect commercial carriers operating in metropolitan areas. Would they experience net savings? Could they pass along cost increases?

Impacts on Transit and Other Modes

Much more needs to be known about possible increases in transit use and ridesharing in specific metropolitan areas. Kain (Vol. 2) notes that "ideally this paper would have presented the results of simulation studies of alternative congestion pricing schemes for three or four representative metropolitan areas, selected to represent different land use patterns, levels of transit use, and congestion levels." Such analyses would provide better estimates of the extent to which congestion pricing would shift demand to carpools and transit and would provide more insight into how transit operators might be able to expand service and how that service might benefit lower-income users.

Modeling

Most regional travel demand and land use forecasting models in use by metropolitan planning organizations (MPOs) do not incorporate a variable reflecting household income, nor do they include price changes as a predictive variable. Thus most MPOs are not in a position to begin estimating the range of consequences that would result from congestion pricing. (Some regional models can be and are being modified to incorporate price signals to some degree.) Simulation of the possible effects of congestion pricing and parking pricing in the Bay Area has been made possible by the previous development of complex models that incorporate price signals to some degree (Harvey, Vol. 2). These models, however, were not designed with the intent of testing pricing strategies. Additional refinements would be desirable. Given the importance of estimating the possible effects of pricing strategies and the variance of these effects with local geographic, demographic, economic, and transportation system characteristics, considerable effort will be needed to develop models applicable in individual metropolitan areas that would incorporate household demand characteristics and be capable of simulating behavioral changes in response to pricing strategies. Such models have been developed in the United Kingdom and in the Netherlands; U.S. models might be patterned on the Bay Area model.

Measures of Congestion

Although it is widely believed that congestion is worsening in metropolitan areas throughout the country, there is little empirical basis to corroborate this impression. Only fairly crude data are available, and these do not measure congestion directly. A few metropolitan areas have consistently measured congestion on specific corridors, but few MPOs have any measures of how congestion has changed over time throughout their regions (Meyer, Vol. 2). Research is needed to develop better congestion measures. Meyer (Vol. 2) outlines several characteristics that such measures might have; most important, they should "have a strong functional relationship to the actual costs of congestion." (Measuring traffic congestion is also reviewed in Appendix A.)

Productivity

Transportation makes up a substantial share of the U.S. economy—in total it accounts for about 17 percent of the gross national product. If congestion pricing strategies became widespread, they would diminish the need for building capacity to serve peak demand and would thereby reduce tax burdens and free more revenues for use in the private economy. Research is needed to estimate the magnitude of these effects and how the efficiency gains of congestion pricing would improve the efficiency and productivity of regions that adopt it.

International Experience

Congestion pricing has sparked interest and policy studies all over the world. Evaluations are under way for London, Cambridge, and Edinburgh, and congestion pricing policies are being considered at the highest levels of government in the Netherlands and Sweden. International colloquia on congestion pricing and other demand management experience abroad would be helpful. Project officials can learn from the design of other projects, U.S. policy makers can learn from the experience of their counterparts, and researchers can learn about the successes and failures of projects in other nations.

Land Use Changes

The relationship between transportation and land use has been difficult to determine. Land use changes occur over long periods of time, and the effects of transportation are difficult to separate from those of other influences. Congestion pricing's effect on land use and urban form is made more difficult to determine by specific policies that private businesses might use to minimize the impact of congestion pricing on their employees. Research is needed on how to measure the short- and long-term relationships between congestion pricing and land use.

Shifting of Burdens

Much of the debate about congestion pricing has focused on how the burdens might fall on certain groups of users and retail and commercial interests. Research on tax burdens and benefits generally has shown that they shift through changes in land costs and labor compensation. The immediate incidence of a tax is not a good measure of how it affects individuals or the economy over time. Research has not been conducted on the longer-term shifts in land, labor, and retail markets. Such analyses would greatly improve the understanding of the net benefits of congestion pricing.

Regional Politics

A successful policy change of a controversial nature at the regional level requires a great deal of public education and constituency building. Congestion pricing faces uncertain political prospects, in part because the benefits are large but diffuse, and the costs, though outweighed by the benefits, are concentrated. Research is needed for specific regions to determine which groups would benefit and which might lose (or even perceive that they might lose) and the influence these groups have at the regional level. A broad class of highway users would save several minutes a day, for example, but they might not be sufficiently motivated by this potential gain to fight for it, and their interests might not be represented at the regional level. Environmental groups might be motivated to fight for the air quality benefits and might be arrayed with groups supporting public transportation. Opposing political interests might include some commercial groups, commercial transportation, and low-income advocacy groups. The opposition, however, might be divided since some, if not many, members within these groups stand to gain. Large employers in a region might be more attracted to congestion pricing than to employee trip reduction programs. Low-income advocacy groups might be attracted to the possible benefits that improved bus services would provide their constituency. Research is needed on whether a successful constituency could be put together that would encompass the broad beneficiaries, how such a constituency might be built, and whether it could be expected to counter opposition. Alternatively, research is needed on whether opposition could be minimized by educating interest groups on how congestion pricing would benefit society generally and even groups within their own constituencies.

Tradable Permits

A new concept for applying congestion pricing would be for the government to distribute "tradable peak driving permits," which individuals or groups could use or sell (Rom, Vol. 2). This proposal was put forward as a way of uniting interest groups and avoiding the political problem of converting a "free" road to a road with a congestion toll. With such a scheme, everyone would retain the privilege of driving during the peak that they now enjoy, and those that did not value this privilege as highly as its market price could sell it to someone who did. Instead of being "tolled off," a user with a low value of time could be "bought off." The concept has obvious appeal for overcoming political problems but should be studied further to determine its feasibility.

REFERENCES

Altshuler, A. 1990. Discussion of C. Winston, How Efficient Is Current Infrastructure Spending and Pricing? In *Is There a Shortfall in Public Capital Investment?* Proceedings of a Conference Sponsored by the Federal Reserve Bank of Boston, Conference Series No. 34, pp. 206–213.

Bhatt, K. Forthcoming. *Congestion Pricing: A Guide for Project Development*. FHWA, U.S. Department of Transportation.

Bishop, G., et al. 1993. A Cost-Effectiveness Study of Carbon Monoxide Emission Reduction Utilizing Remote Sensing. *Journal of the Air and Waste Management Association*, Vol. 43, pp. 978–988, July.

Cambridge Systematics. Forthcoming. *Impact of Urban Congestion on Business* [NCHRP Project 2-17(5)]. TRB, National Research Council, Washington, D.C.

Deakin, E. 1993. Policy Responses in the USA. In *Transport, the Environment, and Sustainable Development* (D. Banister and K. Button, eds.), E. and F.N. Spon, London.

Gomez-Ibanez, J., and K. Small. Forthcoming. *NCHRP Synthesis of Highway Practice: Road Pricing for Congestion Management: A Survey of International Practice*. TRB, National Research Council, Washington, D.C.

Toh, R. 1977. Road Congestion Pricing: the Singapore Experience. *Malayan Economic Review*, Vol. 22, pp. 52–61.

Vickrey, W. 1993. Principles and Applications of Congestion Pricing. *TR News*, Vol. 167, July-August.

Watson, P., and E. Holland. 1978. *Relieving Traffic Congestion: The Singapore Area License Scheme*. Working Paper No. 281. World Bank, Washington, D.C., June, 286 pp.

Weiss, C. 1972. *Evaluation Research: Methods for Assessing Program Effectiveness*. Prentice-Hall.

Wohl, M., and C. Hendrickson. 1984. Some Practical Pricing Problems. In *Transportation Investment and Pricing Principles*, John Wiley and Sons, New York, Chap. 13.

6
Summary

Congestion pricing has great promise: it could reduce congestion significantly while helping to meet air quality and energy conservation goals. Moreover, by relying on a market mechanism, it would accomplish these ends while providing net benefits to society. Congestion pricing, however, has long been the ugly duckling of travel demand management policies. Economists have promised for decades that a swan would emerge if governments would only try it. Transportation officials have consistently thought the policy impractical. Politicians have feared that motorists would pay the fees but hate them, and would then retaliate against the officials who allowed such a duckling into their pond. If this report were being written in the 1960s, 1970s, or 1980s, it would probably end here.

What is different about the 1990s is that congestion pricing proposals are moving forward in the United States, albeit in a cautious and incremental fashion. Why is such an oft-dismissed concept being tried at all? There are several reasons:

- With the passage of the Clean Air Act Amendments in 1990 and the Intermodal Surface Transportation Efficiency Act of 1991 (ISTEA), national transportation policy has shifted toward demand management;
- Congestion has worsened on many routes for many commuters;
- Environmental mandates in federal (and California) law make capacity expansion unlikely in metropolitan areas unable to meet clean air standards;

• Even in areas without air quality problems, cost, other environmental restrictions, and community opposition prevent significant capacity expansions;
• The regulatory restrictions on motorist behavior called for in federal (and California) law appear to some to be less palatable than congestion pricing;
• The substantial revenues that can be raised are appealing in an era of financial stringencies in many states and regions;
• Past efforts to resolve congestion through capacity enhancement have not worked because latent demand fills up any added capacity in areas experiencing population or employment growth; and
• Advances in technology have made it possible to charge users at low cost and with minimal inconvenience or intrusion on privacy.

These reasons why congestion pricing appears more appealing to some do not mean that the political barriers to this policy have disappeared. The political and administrative challenges faced by congestion pricing are as significant as before. Many continue to view congestion pricing as an ugly duckling. What is different is that elected officials in a few places are coming to appreciate congestion pricing's potential to emerge as a swan.

Congestion pricing has a proven track record in stifling latent demand: it has significantly dampened demand for peak-period travel in Singapore since 1975. Peak-period automobile travel into the core of the city almost 20 years later is still some 25 percent below what it was in 1974. No other transportation demand management policy tried in the United States has achieved results anywhere close to these.

Actually, the success of Singapore's policy has not been questioned as much as its transferability to other places. Singapore is not a Western-style democracy, nor do officials there have to deal with the multiple, often conflicting governmental units that exist in U.S. metropolitan areas. The lack of existing institutions to manage regional congestion pricing in the United States remains a significant barrier. The proposals going forward in the United States, however, apply to single corridors in which, despite the conflicting goals of different governmental units within regions, entities (public and private) do exist that can propose and manage the projects.

The kinds of technology that would allow for automatic, convenient pricing are being applied today. The technology being used in the Norwegian toll rings could be readily extended to congestion pricing. Electronic tolling costs far less than attended or automatic toll booths and allows for high traffic volumes since vehicles equipped with transponders do not have to stop at toll booths. Further advances in electronic toll collection, which are being introduced on toll roads in the United States, are making it possible to bill users with minimal intrusion on privacy. The cost of the technology, enforcement, and administration of electronic pricing would total roughly 5 percent of the potential revenues in an areawide application.

Metropolitan areas that are failing to meet federal and state air quality mandates face unattractive choices. The many areas unable to meet federal standards for ozone and carbon monoxide are required by the 1990 Clean Air Act Amendments (CAAA) to propose transportation control measures to reduce the total amount of motor vehicle travel. Employers with more than 100 employees in metropolitan areas classified as in severe nonattainment of CAAA ozone standards must submit plans for employee trip reduction programs by 1994 and be in compliance by 1996. An employer trip reduction policy is already in effect in Southern California; Regulation XV of the South Coast Air Quality Management District requires employers with over 100 employees to develop measures that will increase the average occupancy of the vehicles that employees drive to work. The first-year experience with California's Regulation XV indicates that it did boost average vehicle ridership by 3.4 percent, but this resulted in a net decline in regional travel of, at most, 0.4 percent and cost employers in the region about $150 million. The program is not popular with employers because of the cost and reporting requirements. If travel demand management programs such as employee trip reduction are necessary for meeting air quality goals, they would be more effective if combined with a congestion pricing strategy. Congestion pricing may not be popular initially with motorists, but it is more likely to be effective in improving mobility and meeting air quality goals.

The funding provided by ISTEA for pilot congestion pricing projects has served as a catalyst for the renewed interest in this policy. The federal government has provided incentives, but local government must take the initiative. Although ISTEA allows for up to five pilot projects between 1991 and 1997, only the proposal from the San Francisco Bay Area was selected in the first round of solicitations in 1993. Many other applications were received, but they were not sufficiently developed to be considered projects that might be implemented or they proposed hybrid projects that the federal review panel rejected. This experience reflects the controversy surrounding congestion pricing proposals, the difficulty in proposing eligible projects that are politically feasible, and the considerable gestation period required to develop support at the local level.

Because of the controversy about congestion pricing proposals, careful analyses of how the policy would work at the local level, who would benefit, and how to compensate those disadvantaged by the policy are essential to informing the public deliberations about this policy. The Bay Area proposal benefited by having been subject to extensive modeling efforts to estimate the potential impacts.

POSSIBLE EFFECTS

Travel Behavior

The best available estimates from models, though still approximate, indicate that a $0.06/km ($0.10/mi) charge on all major corridors in the Bay Area and a $0.09/km ($0.15/mi) charge in the greater Los Angeles area would reduce total automobile travel and trips during the peak period by about 10 to 15 percent. The benefits to motorists traveling in congested conditions and willing to pay the fee would be tangible. On average they would save 10 to 15 min per round trip. Society as a whole would gain from the more efficient use of resources.

Emissions

The travel reductions in the Bay Area and in the greater Los Angeles area would result in about a 3 to 9 percent reduction in different vehicular emissions regionwide. (The local effects on some pollutants, particularly carbon monoxide, would be much greater.) These estimates may appear to indicate a small effect on traffic and regional emissions from areawide congestion pricing. Other policies aimed at changing motorist behavior, however, have even smaller effects at the regional level. A combination of reasonably available demand management efforts such as improved transit services, lower transit fares, improved bicycle access, expansion of high-occupancy-vehicle lanes in the Bay Area, and other measures might approximate the effect that congestion pricing alone would have on trips and emissions, but would do so at a much higher cost.

Shifts to Alternative Modes

The estimates of the regional effects of congestion pricing are only approximations. Empirical information is not available for some important potential behavioral responses. Sufficient experience with pricing transportation services indicates that congestion pricing would reduce demand, but the magnitude of that change is not known. In addition, little specifically is known about how motorists might shift the timing of trips; choose alternate routes; choose among solo driving, carpools, and transit options; or simply forgo trips.

Fairness

Lack of understanding about the ability of motorists to adapt to congestion pricing also makes it difficult to estimate the potential hardship on some individuals. The economic effects on commuters of different income levels show that all income groups can benefit from congestion pricing if some of the funds collected are redistributed in ways specifically designed to achieve this goal. As shown in Chapter 4, however, motorists with substantially longer-than-average commutes could be disadvantaged even after revenues are redistributed if they continue to drive alone. Those who do not have better alternatives will drive alone and be made worse off. Those who do have better alternatives than driving alone (who can shift to a carpool or to transit with an acceptable trade-off between lower out-of-pocket costs and time losses) might be made better off because of more attractive carpooling and transit alternatives or higher speeds and, perhaps, more frequent transit service.

Specific Facilities

In contrast to the areawide pricing scheme in Singapore, the toll rings in Norway, and the many studies of regional congestion pricing, the projects moving forward in the United States will apply congestion pricing to individual facilities. (These projects will have to be designed carefully to ensure that diverted traffic does not cause such congestion on alternate routes that the benefits would be lost.) Although these proposals promise considerable congestion relief, the net effects on regional travel, air quality, and energy conservation are likely to be modest. A peak-period toll on the San Francisco–Oakland Bay Bridge, for example, is estimated to reduce precursors to ozone in the region by 0.10 to 0.15 percent. Smaller-scale efforts such as those being proposed and implemented in the United States reflect the controversy and political risk of congestion pricing. Incremental applications may be the only way this policy can be tested in the United States.

POLITICAL FEASIBILITY

Whether congestion pricing will prove politically feasible in more than one or two places remains to be seen. Public and political concerns about fairness and motorist resistance to direct charges for highway use continue to be significant obstacles. The uses of the substantial revenues that congestion pricing can generate provide an opportunity to improve the efficiency of the transportation system, ameliorate the negative impact on adversely affected groups, and result in a net benefit for society. Some individuals would still be hurt, however, and whether they would be more motivated to resist congestion pricing than the majority who would benefit will be demonstrated only in actual practice.

IMPORTANCE OF EVALUATION

Assuming that these early congestion pricing projects are implemented, careful and extensive evaluation is essential. These projects will remain controversial. The quality of the debates about these efforts would be substantially enhanced by reliable information about how traffic flows change, by careful analyses of winners and losers, and by survey research regarding motorist perceptions before and after the change. The early congestion pricing projects are not likely to provide substantial new insight into their potential for relieving regional air pollution and saving energy, but they will provide unique opportunities to learn about motorist sensitivity to price changes, how different groups and individuals are affected, and the political sensitivity to pricing as a transportation policy option. The earliest congestion pricing project to be implemented will be a private toll road in Southern California; this is not an ISTEA congestion pricing pilot project and is not eligible for the extensive evaluation funding permitted for pilot projects. The national interest, however, is sufficient to warrant a substantial federal investment in evaluation of any congestion pricing project. Congestion pricing has the potential to be a useful way to significantly reduce traffic congestion. It is a policy facing substantial political and administrative challenges; overcoming these challenges would be aided with the help of reliable information about the full range of issues surrounding it.

The risks associated with congestion pricing and the nature of policy development in a pluralistic society imply that this policy will progress in small steps. Given that congestion pricing represents a substantial change from the current operation of the road system, such small steps are appropriate. If individual projects succeed, they will help convince policy makers and the public of the benefits of congestion pricing. This process will take time, however; thus it may be many more years before congestion pricing would be applied throughout a metropolitan area in this country. Whether congestion pricing will evolve to this level will depend on how it is implemented, how well it works, and how much motorists and voters come to accept it. Only time, experimentation, and careful evaluation will tell.

Appendix A
Traffic Congestion

According to one measure developed to compare congestion on facilities throughout a metropolitan area, congestion increased by 16 percent between 1982 and 1990 in 50 of the largest U.S. metropolitan areas (Shrank et al. 1993). Congestion in this study was approximated by an index developed by researchers at the Texas Transportation Institute (TTI); as measured by this index, half of the 50 metropolitan areas tracked had reached undesirable levels of congestion by the end of the decade (Table A-1).

This index of metropolitan area congestion, however, is illustrative only. Like all other available measures of congestion based on existing data, the index is an indirect measure of the congestion occurring on metropolitan roadways. In fact, there is no good measure of urban traffic congestion that is comparable across areas and that has been collected consistently over time. Indeed, one other indirect measure of congestion that has been collected over time does not correspond with the upward trend shown by the TTI index (Gordon and Richardson, Vol. 2). As described in more detail below, aggregate trend data on the average duration of the journey to work imply that speeds experienced by the average commuter have not worsened. Gordon and Richardson (Vol. 2) hypothesize that the relative stability in journey-to-work duration indicates that, over time, motorists and businesses make rational choices to avoid delays. These choices include changing location, typically by moving from an urban to suburban or exurban site (Gordon and Richardson,

TABLE A-1 Roadway Congestion Levels in 1989 for 50 Metropolitan Areas (Shrank et al. 1993)

Urban Area	Roadway Congestion Index	Rank	Urban Area	Roadway Congestion Index	Rank
Los Angeles, Calif.	1.55	1	St. Louis, Mo.	0.99	25
Washington, D.C.	1.37	2	Cleveland, Ohio	0.97	27
San Francisco–Oakland, Calif.	1.35	3	Cincinnati, Ohio	0.96	28
Miami, Fla.	1.26	4	Norfolk, Va.	0.96	28
Chicago, Ill.	1.25	5	Austin, Tex.	0.94	30
San Diego, Calif.	1.22	6	Ft. Lauderdale, Fla.	0.94	30
Seattle–Everett, Wash.	1.20	7	Jacksonville, Fla.	0.94	30
San Bernardino–Riverside, Calif.	1.19	8	Albuquerque, N.Mex.	0.93	33
New York, N.Y.	1.14	9	Minneapolis–St. Paul, Minn.	0.93	33
Houston, Tex.	1.12	10	Memphis, Tenn.	0.91	35
New Orleans, La.	1.12	10	Fort Worth, Tex.	0.90	36
Atlanta, Ga.	1.11	12	Hartford, Conn.	0.89	37
Honolulu, Hawaii	1.11	12	Nashville, Tenn.	0.89	37
Detroit, Mich.	1.09	14	San Antonio, Tex.	0.88	39
Portland, Oreg.	1.07	15	Louisville, Ky.	0.86	40
Boston, Mass.	1.06	16	Salt Lake City, Utah	0.85	41
Dallas, Tex.	1.05	17	Columbus, Ohio	0.83	42
Philadelphia, Pa.	1.05	17	Indianapolis, Ind.	0.83	42
Tampa, Fla.	1.05	17	Pittsburgh, Pa.	0.82	44
San Jose, Calif.	1.04	20	Oklahoma City, Okla.	0.79	45
Denver, Colo.	1.03	21	Charlotte, N.C.	0.78	46
Phoenix, Ariz.	1.03	21	El Paso, Tex.	0.74	47
Sacramento, Calif.	1.02	23	Kansas City, Mo.	0.74	47
Baltimore, Md.	1.01	24	Corpus Christi, Tex.	0.72	49
Milwaukee, Wis.	0.99	25	Orlando, Fla.	0.72	49

NOTE: Based on the index used by Shrank et al. (1993), a score of 1 indicates that an area has undesirable levels of areawide congestion.

Vol. 2). Nonetheless, the trends from these two data sources are inconsistent: one implies that congestion is worsening, the other implies that it is not. Which is right?

The apparent inconsistency between the TTI index and trend data on commuting trips can be explained in three ways: (*a*) neither the TTI index nor the journey-to-work data measure congestion directly, (*b*) they are actually measuring different things (traffic volumes on specific facilities versus durations of work trips), and (*c*) the origins or destinations, or both, of work trips are changing over time in response to congestion on facilities. Because of these differences, one would not expect the two trends to be entirely consistent. A brief overview is provided here of these three explanations for the apparent inconsistency between the two measures of congestion.

As background for this review, the broad trends affecting travel demand in metropolitan areas are summarized first. Second, alternative measures of congestion are reviewed. Finally, the reasons for the apparently paradoxical findings between these alternative, but indirect, measures of congestion are summarized.

TRAVEL DEMAND

Total metropolitan area travel in the 1980s grew sharply, which has led to considerable concern about growing traffic congestion. Between 1980 and 1990, vehicle kilometers of travel in urban areas grew almost 50 percent, whereas urban population grew only 12 percent (Gordon and Richardson, Vol. 2). Many metropolitan areas have shown double-digit population growth over the last decade, but even in areas without increased population, more women have joined the work force, job growth has increased, and population and employment growth has continued to shift to outlying areas of metropolises (Pisarski 1987; Gordon and Richardson, Vol. 2).

The amount of traffic growth engendered by these changes has increased faster than population in almost all of the largest metropolitan areas (Table A-2). (The measure of increased traffic growth per lane-kilometer in Table A-2 takes into account the expansions of highway capacity.) Even metropolitan areas with little population growth, such as Chicago, New York, Philadelphia, and Boston, show double-digit increases in traffic as more Americans live in suburban residential areas, work in other parts of the metropolitan area, and increasingly rely on the automobile to travel between them.

TABLE A-2 Trends in Congestion in the 10 Largest Metropolitan Areas (Hanks and Lomax 1990; Shrank et al. 1993)

Metropolitan Area[a]	Congestion Ranking in 1990	Travel Growth, 1982–1990 (% change)	Population Growth, 1982–1990 (% change)	DKT per Lane-Kilometer, 1982–1990 (% change)
New York	9	31.3	0.7	16.5
Los Angeles	1	46.2	15.4	27.2
Chicago	5	49.4	6.1	25.9
Philadelphia	17	48.1	3.7	22.6
Detroit	14	12.1	5.0	−2.4
San Francisco–Oakland	3	47.5	10.5	35.8
Washington, D.C.	2	57.5	27.0	28.1
Boston	16	35.8	3.9	26.1
Houston	10	33.9	19.5	−4.1
San Diego	6	83.6	29.2	61.8

NOTE: DKT = daily vehicle kilometers traveled. 1 km = 0.6 mi.
[a] In order by 1990 population.

Work Trips

During the 1970s and into the 1980s, the suburban rings around metropolitan areas experienced dramatic increases in workers, and the subsequent increases in commuting trips greatly increased peak-period travel (Cervero 1986; Pisarski 1987). Between 1970 and 1980, for example, Houston's population grew 40 percent and its suburban employment mushroomed 90 percent (Pisarski 1987). These employment increases were fueled by the rapid development of office and industrial parks outside urbanized areas—a phenomenon that has occurred all over the country. "Already, two thirds of all American office facilities are in [outlying suburban areas], and 80 percent of them have materialized in the last two decades" (Garreau 1991). The "edge cities" of New Jersey that abridge New York City, for example, have more office space than Manhattan (Garreau 1991).

Rapid job growth in outlying areas of central cities quickly resulted in more jobs in the suburban rings than in the urban core—and in much more dispersed locations. By 1980, commutes from suburb to urban areas—the popular notion of the typical journey-to-work trip—represented only 18 percent of commute trips for metropolitan areas with 500,000 or more population. Suburb-to-suburb trips accounted for 40 percent of commute trips by 1980 and were growing at over twice the rate of suburb-to-central-area trips (Pisarski 1987, 43). These trips are also shorter in duration than other commute trips within metropolitan areas (Gordon and Richardson, Vol. 2, Table 2). The trips may be longer in distance but are completed at higher speeds than commute trips within central cities or commute trips from suburbs to central cities.

Preliminary data from the 1990 census on mode choice and average travel times within individual metropolitan areas indicate a continued strong increase in the choice of the automobile for the journey to work and the growing preference for driving alone—even in highly congested areas like Los Angeles, San Francisco, and Washington, D.C. (Pisarski 1992). The percentage of workers driving alone to work increased from 64.4 percent in 1980 to 73.2 percent in 1990 (DOT 1993). The continued popularity of automobile commuting, even when more time is apparently being spent in congested conditions, traces to several trends. In addition to the increased suburbanization of jobs, other important changes include the continued preferences for the privacy and convenience of automobile travel and lower-density suburban locations for housing (Downs 1992). The declining real cost of gasoline during the last decade has made the choice of automobile travel all the more attractive, as have subsidies that encourage use of the automobile. The deep subsidies that come in the form of employer-provided parking, which is usually costless to the employee, encourage solo driving (Shoup, Vol. 2). The provision of infrastructure through property taxes and other nontransportation revenue sources further masks the true cost of driving automobiles, which thereby encourages their use (Hanson 1992).

Nonwork Trips

The previous discussion of travel demand focused on peak-period travel and work trips. The continued suburbanization of jobs and residences has also coincided with a pronounced increase in nonwork travel (Gordon et al. 1988). Moreover, nonwork trips during the peak are growing much faster than any other category. Between 1977 and 1983, for example, total trips increased by 16.3 percent, but nonwork trips by private automobile in the morning peak period increased by a whopping 42 percent (Gordon et al. 1988). This growth in nonwork peak travel by automobile is most pronounced among suburban residents and is most closely associated with two-worker households and families with children. Most surprisingly, about half of peak-period travel is now made up of nonwork trips.[1]

Summary

Underlying the sharp increases in travel demand in recent years are some important trends. The entry of women into the work force has caused travel demand to outpace population growth. The suburbanization of jobs and continued suburbanization of residences have caused metropolitan

[1] As argued by Gordon et al. (1988), given that more than half of peak-period trips are not work trips, the case for congestion pricing is all the stronger. Presumably the demand for nonwork trips is more elastic than that for work trips. Hence a substantial fraction of trips could be shifted with modest prices. Although this argument is quite plausible, it is not clear whether peak-period nonwork trips are actually occurring on the most congested facilities likely to be subjected to congestion pricing. For example, Giuliano (Vol. 2) notes from her survey research on very congested facilities that most of the travelers are indeed work travelers. Nonetheless, even if only some of the peak-period travelers on congested facilities are making discretionary trips, pricing would most certainly alter their behavior.

areas to spread. Suburbanization has caused a dispersal of automobile travel over a wider, less dense area in which the automobile is the only convenient alternative for work trips. A much faster rate of growth has occurred in suburb-to-suburb trips than in suburb-to-central-area trips. Substantial growth has occurred in nonwork trips during the peak period.

MEASURING TRAFFIC CONGESTION

Although any motorist is aware of traffic congestion when he or she experiences it, available data to provide a precise measure of congestion are another matter. Available data provide only indirect measures of congestion. The alternatives are to infer congestion from (*a*) traffic volumes on facilities or (*b*) changes in trip durations over time.

Facility-Based Measure

The *Highway Capacity Manual*, the most widely used reference in highway planning and design, does not define congestion on facilities explicitly (TRB 1985). Rather, it lays out a definition of *capacity*, which must be understood relative to the desired or expected *level of service* (LOS). The *Highway Capacity Manual* defines the maximum capacity of a facility as "the maximum hourly rate at which persons or vehicles can reasonably be expected to traverse a point or uniform section of a lane or roadway during a given time period under prevailing roadway, traffic, and control conditions." LOS is defined as "a qualitative measure describing operational conditions within a traffic stream, and their perception by motorists and/or passengers." Six levels of service are described, from LOS A, which is free flow, to LOS F, which is characterized by stop-and-go traffic (see accompanying text box).

The definition of congestion for any facility using the LOS approach depends on the quality of service "expected," which may vary between designers and users and even among users. In measuring congestion with the LOS approach, the judgment of the designer or analyst plays a large part in what is defined as congestion.

Hanks and Lomax (1990), researchers at TTI, developed a widely cited index estimating metropolitan area congestion that is based on such judgments. The index is one of the few measures that can be consistently applied across metropolitan areas (Hanks and Lomax 1990; Shrank et al.

LEVELS OF SERVICE

A: Free flow. Drivers unaffected by others in traffic stream, free to select speed and maneuver.

B: Still free flow, but with more notice of other drivers. Slight loss in freedom to maneuver.

C: Stable flow, but the beginning of the flow in which operations become significantly affected by the presence of other drivers. Declining comfort and convenience.

D: High density but stable flow. Maneuvering restricted.

E: At or near capacity. Speeds reduced to slow but uniform flow. Maneuvering extremely difficult. Flow unstable.

F: Stop-and-go flow. Queues develop behind breakdowns in flow.

1993).[2] This index estimates congestion on the freeways and principal arterials within an urbanized area by comparing daily traffic per lane-kilometer with a judgment regarding the traffic level at which congestion begins.

The estimated congestion is weighted by total travel on each system such that a value of 1.0 indicates the beginning of congestion for an urban area. The TTI roadway congestion index (RCI) is (1 mi = 1.6 km)

$$\mathrm{RCI} = \frac{\left(\frac{\text{Freeway}}{\text{VMT/ln-mi}} \times \frac{\text{Freeway}}{\text{VMT}}\right) + \left(\frac{\text{Principal Arterial Street}}{\text{VMT/ln-mi}} \times \frac{\text{Principal Arterial Street}}{\text{VMT}}\right)}{\left(13{,}000 \times \frac{\text{Freeway}}{\text{VMT}}\right) + \left(5{,}000 \times \frac{\text{Principal Arterial Street}}{\text{VMT}}\right)}$$

where VMT is vehicle miles traveled.

Hanks and Lomax base their estimate of congestion on the proportion of traffic volumes on freeways and principal urban arterials that exceeds LOS C (Hanks and Lomax 1990, A.7-A.11). According to their analysis of

[2] The congestion index is updated annually. Shrank et al. (1993) provide 1990 data for 50 urban areas.

traffic volumes and delays on Houston-area roads over the 1970 to 1985 period,

> The maximum freeway service flow rate for level-of-service C is 1,550 passenger cars per lane hour (volume/capacity ratio equal to .77 for a 70 mph design speed facility) [(TRB 1985)]. Using average values for K factor (the percentage of daily traffic volume during the peak hour) and directional distribution, and including some adjustment for trucks, these values can be interpreted to indicate that 15,000 vehicles per lane day is an estimate of the beginning of level-of-service D operation. . . . The use of the boundary between level-of-service C and D as the beginning of congestion is consistent with reports by the Department of Transportation to Congress on the status of highways in the United States (congestion begins at a volume/capacity ratio of .8) and the AASHTO Policy on Geometric Design of Highways and Streets. (Hanks and Lomax 1990)

Although their analysis leads Hanks and Lomax to define 15,000 vehicles per lane-day for freeways as a point at which congestion is becoming critical, in their index they use 13,000. They define 13,000 vehicles per lane-day as "a measure of approaching congestion," which is consistent with a measure used in other FHWA and Texas Department of Transportation reports.

The TTI index is based on counts of daily traffic volumes rather than measures of peak-period traffic volumes because the daily traffic count on most major routes is an available statistic, whereas peak-period traffic volume is not. The selection of daily traffic levels at which congestion occurs is based on estimates of the average percentage of travel occurring in peak periods. Traffic congestion, however, is specific to individual routes, corridors, or intersections. The use of average peaks gives, at best, an indirect gauge of the congestion that motorists encounter on their trips. Moreover, as noted above, the percentage of work and nonwork traffic occurring during peak periods appears to be changing over time, with peak-period traffic growing faster (Gordon et al. 1988). As a final point, the aggregation of the traffic data to a metropolitan-wide average, although useful as an indicator for the metropolitan area, may mask conditions that are much worse on some facilities and much better on others. Despite these weaknesses, the increases in the TTI measure of metropolitan congestion over time are plausible because the index is based on measures of traffic volumes compared with estimates of highway capacity, which is not increasing nearly as fast as traffic. As long as these conditions continue, congestion must increase as demand approaches capacity.

Trip-Based Measure

Gordon and Richardson (Vol. 2) compared survey data on average work trips from different national surveys with the areas ranked most congested by facility-based measures. Their analysis shows no correlation between the average duration of the journey to work and congestion in the cities Hanks and Lomax identified as the most congested (see Gordon and Richardson, Vol. 2, Table 1). Los Angeles, for example, which is ranked first by the TTI index, is placed in the middle of metropolitan areas on the basis of average work trip times. Hartford, which ranks at the bottom of the TTI list, has almost the longest average work trip times of the areas studied. Moreover, despite the increased congestion levels over time indicated by the TTI index, the average work trip duration between 1983 and 1990 has hardly changed (Gordon and Richardson, Vol. 2, Table 4).

Gordon et al. characterize the disparity between these findings this way:

> The commuting paradox reflects the apparent contradiction between perceptions of worsening traffic congestion and evidence of either declining or stable commuting times. However, not only is there no contradiction but the two phenomena are causally related. Rational commuters will, sooner or later, seek to escape congestion by changing the location of their homes and/or their jobs. This type of adjustment is easier to make in large, dispersed metropolitan areas with alternate employment subcenters and a wide variety of residential neighborhoods. The process is facilitated by the decentralizing location decisions of firms seeking to move closer to suburban labor pools. (Gordon et al. 1991)

Although it is plausible that congestion brings about the suburbanization of jobs and residences, it may be premature to state that average commute trips are not increasing. The estimates of travel times used by Gordon and Richardson are based on surveys in which respondents are asked to estimate the duration of their work trips. As the authors acknowledge, one would expect that respondents would round off trip times in 5-min increments. Thus it would be difficult to discern small changes in the average over time. Moreover, respondents may have difficulty accounting for the increased variability of trips. In other words, respondents may be encountering congestion more frequently than in the past. Although the average commute on a "good" day may still be, say, 20 min, the frequency of "bad" days may be increasing. The subjective measure of average trip times would tend to make it difficult to account for such changes.

An additional shortcoming of this measure is that in order to develop an average commute time for the entire metropolitan area, these estimated averages are then averaged again. In short, this is a highly aggregated measure of a phenomenon that is localized and specific. Many travelers within a metropolitan area may be experiencing sharply increased congestion, even though in aggregate these time losses are being offset by the growth in shorter durations in expanding parts of the same area.

There is also a problem of discrepancies among individual surveys. The national census data relied on by Gordon and Richardson, for example, do not correspond with recently released local travel surveys of the Los Angeles and San Francisco metropolitan areas. Preliminary census data for Los Angeles County report an average commute time of about 27 min, but a recent survey by the Southern California Association of Governments (SCAG) reports an average for 1991 of 29.2 min (Pisarski 1992; SCAG 1993). For the San Francisco Bay Area, the discrepancies are much larger. In the Bay Area's 1981 survey, average commutes were 9 percent higher than those reported to the census, and in 1990 the average commutes were 18 percent higher (Purvis 1994). In contrast to the relative stability in average commute times reported in census data, local surveys for both Los Angeles and the San Francisco Bay Area over a 20- to 30-year period indicate that reported average commutes are increasing. Between 1967 and 1991, average commutes in Los Angeles County increased by 21 percent, and between 1960 and 1990, average commutes in the San Francisco Bay Area increased 17 percent (SCAG 1993; Purvis 1994).

Despite these data discrepancies, the argument that Gordon and Richardson make—that individuals and businesses relocate to avoid congestion—has partial corroboration from other sources. Other data cited above indicate the substantial increase in the suburbanization of jobs during the 1980s and the continued faster growth in suburban and exurban population than in central-area population. Thus there is supporting evidence that changes in jobs and residences within metropolitan areas are altering the commuting behavior of many workers. Whether these changes are caused by or are large enough to offset growing congestion, however, is not certain.

SUMMARY

Two different indirect measures of congestion appear to tell different stories. One indicates that areawide congestion is worsening; the other indicates that behavioral changes keep congestion from worsening because people make rational choices to avoid delay. It can be argued that both stories are true. Consider this example. Drivers become frustrated with the length of time spent on a congested route to work; over time they change their residence to shorten the commute time. [Some may choose to move closer to work in order to reduce travel times; some may choose to move both job and residence to outlying areas to escape congestion (Wachs 1993).] Despite the change to a different route for most of the trip, the congestion on that facility might increase because of increased trip making by others. The time savings may offset the growing time losses of others using the increasingly congested facility. Thus it is possible to have both increased congestion on facilities and a stability in average commute times.

On the other hand, both of the indirect measures of congestion reviewed here have weaknesses. Both are highly aggregated when congestion is local and specific. Moreover, the facility-based measure assumes average peaking characteristics that, in practice, would tend to vary widely across routes and that may be changing over time. The trip-based measure could have biases because of its subjective nature, and there is some unexplained discrepancy with other data sources.

In the context of the issues being discussed in this report, whether congestion on facilities is increasing or whether congestion is being held in check by locational changes is almost a moot point. Traffic congestion is an important social problem because it wastes time and consumes resources that could be invested more productively. The theoretical underpinnings of pricing to reduce congestion are outlined in Appendix B.

REFERENCES

ABBREVIATIONS

DOT	U.S. Department of Transportation
SCAG	Southern California Association of Governments

Cervero, R. 1986. *Suburban Gridlock.* Center for Urban Policy Research. Rutgers University, Brunswick, N.J.

DOT. 1993. *Journey-to-Work Trends in the United States and Its Major Metropolitan Areas: 1960 to 1990.* Office of Highway Information Management, Federal Highway Administration.

Downs, A. 1992. *Stuck in Traffic: Coping with Peak-Hour Traffic Congestion.* The Brookings Institution, Washington, D.C.

Garreau, J. 1991. *Edge City: Life on the New Frontier.* Doubleday, New York. 546 pp.

Gordon, P., et al. 1988. Beyond the Journey to Work. *Transportation Research*, Vol. 22A, No. 6, pp. 419–426.

Gordon, P., et al. 1991. The Commuting Paradox. *Journal of the American Planning Association*, Vol. 57, No. 4, Autumn, pp. 416–420.

Hanks, J., and T. Lomax. 1990. *Roadway Congestion in Major Urban Areas, 1982 to 1988*. Report 1131-3. Texas Transportation Institute.

Hanson, M. 1992. Automobile Subsidies and Land Use: Estimates and Policy Responses. *Journal of the American Planning Association*, Vol. 58, No. 1, Winter, pp. 60–71.

Pisarski, A. 1987. *Commuting in America: A National Report on Commuting Patterns and Trends*. Eno Foundation for Transportation, Inc., Westport, Conn.

Pisarski, A. 1992. *New Perspectives in Commuting Based on Early Data from the 1990 Decennial Census and the 1990 Nationwide Personal Transportation Study (NPTS)*. U.S. Department of Transportation, July.

Purvis, C. 1994. Changes in Regional Travel Characteristics and Travel Time Budgets in the San Francisco Bay Area: 1960–1990. To be published in *Transportation Research Record*, TRB, National Research Council, Washington, D.C.

SCAG. 1993. *Summary Findings: 1991 Southern California Origin-Destination Survey*. February.

Shrank, D., et al. 1993. *Estimates of Urban Roadway Congestion, 1990*. Research Report 1131-5. Texas Transportation Institute.

Transportation Research Board. 1985. *Special Report 209: Highway Capacity Manual*. National Research Council, Washington, D.C.

Wachs, M. 1993. Learning from Los Angeles: Transport, Urban Form, and Air Quality. *Transportation*, Vol. 20, No. 4, pp. 329–354.

Appendix B

Theory, Experience, and Estimated Effects

In 1920, long before hundreds of thousands of motorists were faced with the daily frustration of congested routes like the Santa Monica Freeway or Shirley Highway, economists developed the basic theory of road pricing (Pigou 1920; Knight 1924). In its simplest formulation, road pricing would require motorists using the roadway to pay for the congestion they cause other motorists. In the short run the optimal toll would reduce congestion to its most efficient level. Congestion pricing theory holds that congestion tolls not only would optimize the use of the current road system and generate substantial net savings in the short run, but in the long run, and if other assumptions hold, they also would generate just enough revenues to provide for demand in the future (Mohring and Harwitz 1962).

In the 1960s, some four decades after the theory was developed, economists began suggesting that the theory be applied to solve the growing congestion problem (Vickrey 1959; Walters 1961). In 1975, Singapore imposed the first congestion pricing system. The pros and cons of Singapore's congestion pricing approach, which has been praised for its traffic constraint but criticized as a model for congestion pricing, are described in the second section of this appendix.

Congestion pricing has been proposed in the United States before. These proposals, however, did not advance very far. Despite fairly compelling arguments in favor of congesting pricing from economic theory, political opposition caused by concerns about the potential effects on the poor, opposition by downtown businesses, and resistance to the concept of charging motorists directly for the use of roads has stymied the progress of such proposals.

In the first section of this appendix, the theory of congestion pricing is summarized. In the second section, the experience with congestion pricing abroad is reviewed, and in the third, estimates are provided of the possible effects of congestion pricing if applied in U.S. metropolitan areas.

THEORY

An Overview

The traditional approach to resolving congestion problems has been to expand capacity by adding new lanes or building new roads. This approach is problematic for both obvious and subtle reasons. Among the more obvious reasons, *(a)* in most urban areas addition of more lanes would be very costly because of the high cost of real estate, especially at a time when governmental agencies at all levels are short of funds, and *(b)* in some areas capacity cannot be enhanced because of concerns about air pollution or community opposition, or both. Among the more subtle reasons is the recognition that building new capacity induces new demand or shifts in existing demand that soon congest the new facility (Downs 1962).

Economists offer an alternative to building our way out of congestion that would instead change the behavior of some road users. They observe that whenever the price of using some scarce, valued good does not increase as demand increases, that good will be in short supply. Shortages will be acute if supply cannot be readily enhanced. This is typical of goods in industries with high capital or fixed costs. Throughout the economy, when demand for some commodity or service exceeds supply, the price tends to rise until demand and supply are in balance.

The basic theory of congestion pricing for roads has changed little since Knight's (1924) formulation.[1] Speed–traffic flow curves plotted by traffic engineers indicate that as the volume of traffic on a road approaches design capacity, speeds and traffic flow decline sharply (Figure B-1). Once capacity has been reached, the addition of motorists into the traffic stream causes the flow of vehicles per lane per unit of time to decrease, resulting in the

[1] This discussion draws heavily from work by Morrison (1986) and Hau (1992b). Hau provides an extensive, nonmathematical treatment of congestion pricing theory.

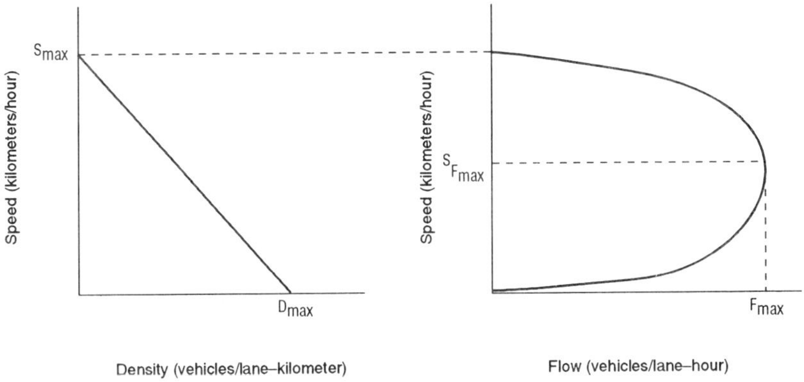

FIGURE B-1 Derivation of the speed-flow curve (Morrison 1986).

backward-bending curve in Figure B-1.[2] A general interpretation may clarify this relationship. Consider a road with a design speed of 97 km/hr (60 mph) on which traffic is moving at an average speed of 80 km/hr (50 mph). At that speed, a traffic lane would accommodate 900 to 1,000 vehicles per hour. As traffic increases, speeds might fall to 56 km/hr (35 mph), but the lane can actually carry more traffic each hour—nearly 1,900 passenger cars—because at slower speeds the space between vehicles is reduced. As capacity is reached, however, the lane starts to become so crowded that speeds fall below 56 km/hr and capacity diminishes sharply. At very high volumes, the risk that traffic flow will degenerate to stop-and-go conditions is quite high.

The shape of the speed-flow curve as demand approaches capacity provides a key clue to the theory of congestion pricing. This theory can be illustrated by examining how delay increases as the flow of vehicles per lane-hour increases. The speed-flow curve shown in Figure B-1 can be inverted from kilometers per hour, which measures speed, to hours per kilometer, which measures trip duration (Figure B-2a). Speed is at its maximum in low traffic volume; hence delay is at its minimum. But as the number of drivers in a traffic stream increases, the average delay at each level of flow increases. This is shown by the curve labeled *AVD* in Figure B-2a. Because the average delay at each level of flow is increasing, it follows that the contribution to delay by each additional driver is increas-

[2] Although the backward-bending portion of the curve is shown as a smooth parabola, traffic engineers consider the entire area highly unstable. Once the volume is at or near capacity, any interruption can cause significant declines in flow, resulting in a rapid transition from peak flow to stop-and-go traffic (Transportation Research Board 1985).

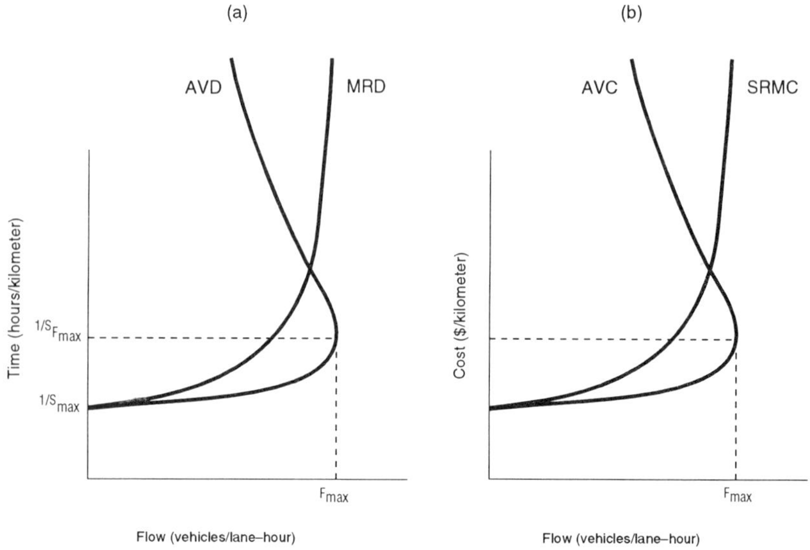

FIGURE B-2 Derivation of the cost-flow relationship (Morrison 1986). (*a*) Trip duration; (*b*) trip cost.

ing even more; this marginal delay is illustrated by the curve labeled *MRD* in Figure B-2a. The marginal delay is the increment in delay for each incremental increase in traffic volume. As can be seen by examining these two curves, each additional driver's contribution to delay is much larger than the increased average delay that he or she experiences (the marginal delay curve is much higher than the average delay curve and is rising faster).

The average delay can be converted to a dollar cost by multiplying it by the average value of time. After variable operating costs are added to cover fuel and maintenance, it becomes an estimate of short-run average variable cost (*AVC* in Figure B-2b). Similarly, the marginal delay curve is converted to a short-run marginal cost (*SRMC*) curve, which gives the cost to all drivers of adding one more driver to the traffic stream during the same time period (Figure B-2b). As with the delay curves, these two cost curves are closely related and based on the same information, but are computed differently. Average variable costs are the total variable costs (*TVC*) per unit of flow at each level of traffic flow (*F*) (or *TVC*/*F*) and the short-run marginal cost is equal to the change in total variable costs with each change in traffic flow $[(TVC_1 - TVC_2)/(F_1 - F_2)]$.

In the absence of a congestion toll, the quantity of traffic flow will be that which occurs at the point where the average cost curve intersects the demand curve (Figure B-3). As can be seen, marginal costs are considerably higher than average costs at this point. Additional drivers joining a congested traffic stream, absent a congestion toll, may only be aware of the average cost they will experience and are largely unaware of the increased costs they are imposing on all other drivers.

The theory of congestion pricing has been aptly described by Mohring and Anderson (1994):

> Urban travelers both experience congestion and contribute to it. Inducing the operator of a vehicle—any vehicle—to remove it from a traffic stream would save not just its occupants' own time but also the time cost they would otherwise impose on other travelers by adding to the road's congestion level. This time cost is the change in time per trip the departed vehicle would have produced *times* the number of vehicles that change would have affected.

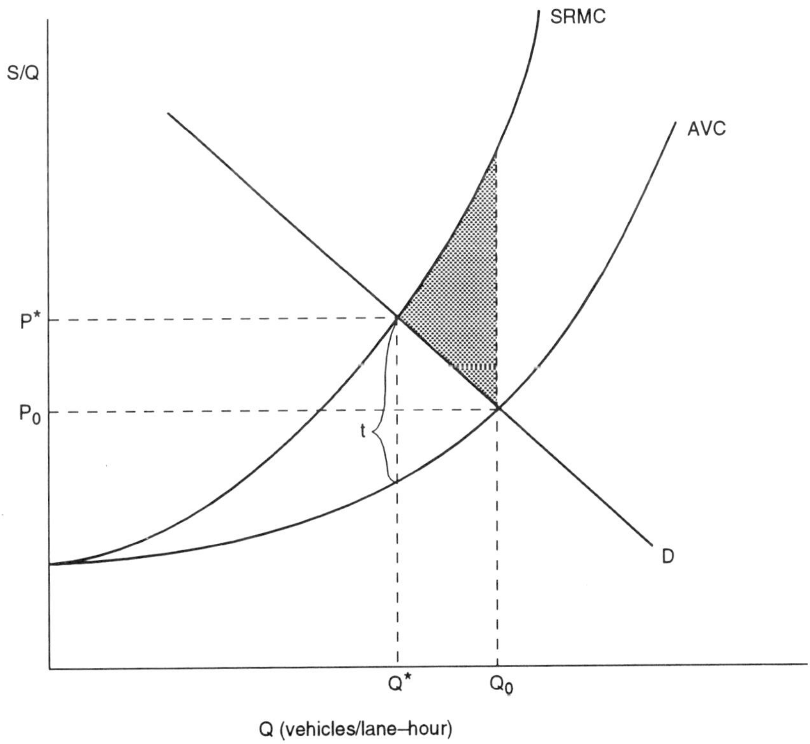

FIGURE B-3 Optimal congestion toll and welfare loss (Morrison 1986).

A commonly invoked rule of safe road behavior is that drivers should stay one car length behind the vehicles they follow for each ten miles an hour of travel speed. If all travelers follow this rule and all would travel at 60 miles per hour (i.e., would take one minute to travel a mile) on an otherwise unused expressway, there would result the relationships between the instantaneous ratio of actual traffic volumes to "ideal" capacity (about 2,000 vehicles per lane-hour on an expressway) and the average (AC_1) and marginal (MC_1) travel times per mile that are given by the solid curves in Figure 1. Curve AC_1 depicts the travel-time costs that individual

Figure 1. Relationships bertween Volume/Capacity Ratios and Travel Time

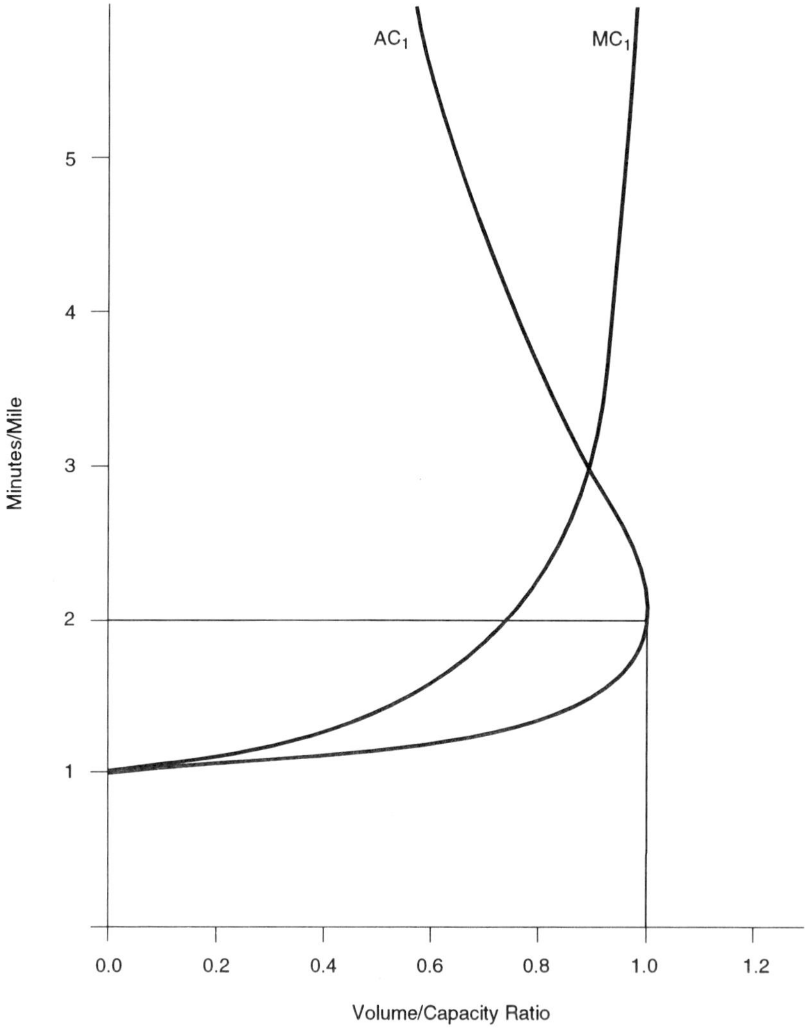

travelers directly experience; curve MC_1 includes these costs plus those that each vehicle operator imposes on others by adding to congestion. Curve AC_1 illustrates a commonly observed phenomenon of urban-expressway travel: maximum traffic flow occurs at about 30 mph. Above 30 mph, lower speeds result in increased traffic flows. In the top, backward-bending portion of AC_1 where it takes more than two minutes a mile, however, further speed reductions lower traffic flows.

Peak-period travel does not take place at a constant rate but, rather, gradually increases to a peak then decreases. Someone traveling at the peak of the peak period is more likely to experience the backward-bending portion of curve AC_1 than is someone who travels at the beginning or end of the peak. Still, both peak-of-the-peak and fringe-of-the-peak travelers do almost always get where they are going; a peak-of-the-peak trip just takes longer. We have, therefore, used marginal and average travel-time relationships similar to those given by the dashed curves, AC_2 and MC_2 in Figure 1 in deriving the toll estimates that are reported in Section V.[6]

[6] The formulae for the solid and dashed average cost curves are, respectively, $N/K = 4(1 - t^*/t)t^*/t$ and $t = 1 + (N/K)^4$ where N/K denotes the ratio of vehicle volume to ideal capacity and t and t^* are respectively actual travel time per mile and travel time per mile at a zero N/K ratio.

The marginal delay cost imposed on other drivers is made up of costs that are external to a motorist's choices. The existence of external costs in a transaction indicates that decisions are being made without accounting for their full effects on others. When highway capacity is allocated according to the average cost, the total loss to society due to these external costs is simply the sum of all delay costs in excess of the most efficient level. This loss is represented by the shaded area above the demand curve between the average and marginal cost curves (Figure B-3). This shaded area represents the excess delay imposed on all motorists because of the consumption of highway capacity by drivers in excess of their willingness to pay if charged the marginal cost.

The optimal traffic flow would occur at the point where the short-run marginal cost curve intersects demand, but without somehow making drivers internalize the external costs they impose on others, the optimal flow will not occur. The optimal flow would be achieved by requiring each additional driver to, in effect, compensate all others for the delay he or she causes by joining the traffic stream. This toll (t in Figure B-3) would equal the difference between the average variable cost and the marginal cost at that point.

When motorists are faced with a toll that represents the full social cost of using a congested highway, those who value the use of the less-congested road at that time will pay the toll and save time. Those who place less value on the use of the highway at that time will make other choices, such as delaying their trip, traveling on another route, traveling by transit, or deferring the trip. In the short run, defined as that period during which capacity is fixed, the time savings that result from the imposition of the optimal toll makes society as a whole better off.

Over the long run, the willingness of travelers to pay the full marginal cost would give a better indication of the demand for new capacity. It can be shown that pricing highway capacity investment needs according to their marginal cost will generate revenues just sufficient to cover the cost of providing new facilities (if other assumptions discussed below hold) (Mohring and Harwitz 1962; Keeler and Small 1977).

Theoretical and Practical Challenges

Although theory provides a compelling case for congestion pricing, it has been challenged on at least three fronts.[3] As described in the following three sections, (*a*) the assumptions incorporated by theory have been questioned, (*b*) concerns have been raised about the effects on low-income road users, and (*c*) concerns have been raised that the pricing system adopted in practice will not meet the efficient pricing criteria assumed by economists. Regarding the last concern, even one of the strongest proponents of congestion pricing has noted that "experience with the pricing of public services is not such to give confidence that in practice a close approach to an efficient optimum can be achieved" (W. Vickrey, unpublished paper, 1992).

Questions About Assumptions

As indicated in the previous section, congestion pricing has a strong theoretical rationale for maximizing the efficient use of existing capacity. In addition to its short-run benefits, marginal cost congestion pricing

[3] Hau (1992b) discusses the full range of theoretical challenges to congestion pricing.

should also lead to an optimal allocation of resources in the long run. This additional advantage has been questioned. It can be challenged whether an optimal allocation of resources will occur with congestion pricing because of potential violations to three assumptions: perfect competition, no scale economies, and a specified income distribution. As described below, there are practical responses to the criticisms that these assumptions are not perfectly met. Moreover, it has been shown that whereas the conditions listed are sufficient conditions for marginal cost pricing to be optimal, they are not necessary (Ng 1977).

One argument against marginal cost pricing optimally allocating resources in the long run comes from the theory of the "second best"; that is, market imperfections in aspects of the economy not governed by market pricing can still lead to inappropriate price signals and less-than-optimal resource allocation. The validity of this argument in the case of urban transportation depends on the degree to which pricing in the economy approximates the competitive norm assumed by microeconomic theory and on the existence of substitutes for transportation. The former argument could apply in limited sectors, but Meyer et al. (1965) judge the economy as a whole to be fairly competitive and conclude that substitutes for transportation seem to be a limited concern. Though they acknowledge that empirical evidence is less than compelling, Meyer et al. suggest that "second best" problems are easily exaggerated and conclude that adoption of marginal cost pricing in transportation would probably improve the general welfare.

In the long run, in the case when the supply of highway capacity is not fixed, theory also predicts that an optimum toll may be just sufficient to cover the cost of providing the facility. This outcome, however, is more complicated because it depends on the assumption that returns to scale are constant. If returns to scale were increasing rather than constant, for example, congestion tolls would not generate sufficient funds to cover the cost of the facility. If returns were decreasing, marginal cost pricing would generate more-than-adequate revenues. In the latter case, if supply is not enhanced, a long-run income transfer will take place. Either of these two outcomes would violate the optimal resource goal of congestion pricing. Hau's (1992b) review of the studies examining the issue of returns to scale in highway construction indicates that evidence is available on both sides; he concludes that surpluses and deficits would coexist. Other economists (Meyer and Gomez-Ibanez 1981) conclude that economies of scale probably offset diseconomies of scale. Regardless of the existence of scale economies in the long run, however, short-run congestion pricing will still generate the most efficient use of the existing capacity.

Hau (1992b), among others, argues that the revenues earned from congestion pricing should be reinvested in the transportation sector to appropriately respond to transport demand. Diversion of the funds to nontransportation uses may help compensate groups initially opposed to congestion pricing but could also result in less efficient expenditures on transportation facilities. The use of the revenues is a key issue in developing a congestion pricing proposal that satisfies economic criteria.

However, that congestion tolls cover capital and operating costs on an "optimally designed, constant-returns-to-scale road network" does not mean that all the revenues should be spent on capacity. In the United States most capital costs for highways have been paid on a "pay as you go" basis. Thus revenues earned from pricing such highways need not be earmarked for additional investments in capacity.

Concerns About Regressivity

Congestion pricing is often advocated as a fairer method of providing for highway capacity than the current user tax system because it requires those motorists who wish to use the capacity during the peak period to pay a premium. A limitation on the improved fairness of congestion pricing is that motorists have different incomes and different abilities to pay (although it should be noted that the existing sources of highway finance, mostly gasoline and property taxes, along with sales taxes in some areas, are also regressive).[4] In economic welfare theory, the lack of equally distributed impacts is dealt with by making payments to those disadvantaged by a change in policy from the revenues generated by the policy change. In this case, lower-income highway users could be compensated by using some of the revenues earned from congestion pricing. Although the adverse impact on the poor as a group can be mitigated with side payments, compensating all adversely affected *individuals* would defeat the purpose of congestion pricing, since its success depends on the shift by some users to other modes or times for their trips. Compensation to adversely affected *groups* could be provided, however, without undermining the effect of congestion pricing on demand. But the nature of this compensation is central to achieving the

[4] Of course, in the private economy air travelers and restaurant customers also have different abilities to pay, but there is little social concern about charging for these services on the basis of peak demand rather than ability to pay.

welfare gain. In the short run, Small (1983) and Hau (1992b) have shown that before the funds earned from congestion pricing are reinvested, the majority of highway users are made worse off by congestion tolls. Judicious uses of the revenues can reverse this loss.

Practical Concerns

Many students of congestion pricing have discussed the difficulty of valuing social costs (in this case the value of time) and in gaining acceptance of any estimate (Zettel and Carll 1964; Meyer et al. 1965; Altshuler 1979). This lack of certainty makes estimates of congestion tolls somewhat speculative, but some practical solutions may be available. Despite lack of consensus among analysts, estimates of the value of time are routinely used in developing the benefit estimates that underlie transportation construction projects. In addition, Wohl and Hendrickson (1984) suggest that experimenting with alternative congestion prices until the market is cleared is the simplest way of finding how society values scarce road capacity. This solution, however, is complicated by two factors. Given the problems associated with increasing transit fees and tolls, it may be unrealistic to expect public authorities to be able to experiment with prices. Second, congestion pricing is meant to reduce congestion to its most efficient level, not eliminate it. Because of the vagaries of weather and the unpredictability of some factors influencing demand for peak-period travel, some congestion will occur. The cost of trying to respond to all such influences such that congestion did not occur is likely to exceed the benefits. Hence there is a point at which a certain degree of congestion is efficient. Determining when the efficient level of congestion has been obtained, however, may be a subjective and controversial enterprise.

Whether the compromises necessary in the real world would reduce or vitiate the potential benefits raises fundamental concerns. Vickrey (1959; unpublished paper, 1992), for example, has argued that an efficient pricing system would require smooth changes in price from the peak to the off peak (and vice versa) and should vary with changes in actual traffic conditions. In this scenario, drivers with different values of time would shift their travel times to match the toll they were willing to pay. This scenario would also reduce the problem associated with abrupt changes in the fee. When abrupt changes occur, such as when daily restrictions on high-occupancy-vehicle (HOV) lanes are eliminated, motorists are given an incentive to try to game the system by queueing up on connecting routes (and often blocking traffic) to take advantage of the change. A practical problem with varying tolls is that if they truly vary with demand, they would have to vary with some frequency; Vickrey (unpublished paper, 1992) argues that such variations are required to account for frequent variations in weather and unique situations such as sporting events.

Else (1986) has used queueing theory to illustrate the requirement for varying prices, but his concern is somewhat different. He points out that under conditions of hypercongestion, the congestion caused by a driver (the marginal social cost imposed by the driver) depends on the amount of following traffic. If there is little following traffic behind a bottleneck, relatively less burden is imposed once the bottleneck has been opened than if there is a steady stream of traffic behind the bottleneck. This leads Else to conclude that the fee at the beginning of a peak may need to be larger than that at the end. Vickrey (1959; unpublished paper, 1992) considers the same point and concludes that efficient pricing may require charging drivers after the amount of congestion they cause has been determined.[5]

Such variable prices, however, make it difficult for drivers to have an accurate foreknowledge of the price they will encounter. Without a clear price signal to drivers well in advance of their commitment to traveling on a certain route at a certain time, the demand-smoothing effect predicted by Vickrey may not occur. On the other hand, without varying charges to account for unique or variable conditions, circumstances cannot be avoided that would result in periods of both congestion and underutilization. As a result, the types of congestion pricing imposed in practice, though potentially more efficient than the current system of relying on average delay cost to ration demand, may not reach the optimal results predicted by theory (Gomez-Ibanez 1992).

EXPERIENCE

Whereas the theory of congestion pricing sounds intriguing, many have doubted that congestion pricing would be practical. Experience in Singapore indicates that congestion pricing significantly affects traffic flow. Past

[5] A pricing system that approximates this kind of charging is under development in Cambridge, England. The local government is developing a proposal that would charge motorists with the actual level of congestion they experience. Vehicles and the roads in a heavily traveled section of the city would be outfitted with electronic equipment that would provide a "real time" measure of congestion. Each vehicle would have a meter that would record the charge. The Cambridge proposal, however, would still not account for whether the trip was near the beginning or the end of the peak period.

proposals for congestion pricing in the United States, however, have not advanced very far.

Singapore

In 1975, the government of Singapore imposed the first congestion pricing system (Button and Pearman 1985, 39–46).[6] This system, along with the introduction of congestion pricing on Route A1 south of Paris in 1992, are the only examples of roadway congestion pricing.[7] In 1975, a 6.2-km^2 (2.4-mi^2) area in the central city was marked off, and any motorist wishing to enter this area during the morning peak period had to display a windshield sticker, which cost approximately U.S.$1.65 per day initially. Stickers could be purchased from post offices and road-side booths. The fee was enforced by officers at about 30 entry points to the zone. Other policies were also implemented at roughly the same time to discourage automobile ownership and use. Taxes on vehicle ownership and parking rates in the downtown area were increased sharply. To encourage more transit use, park-and-ride lots (10,000 spaces) were provided on the periphery of the downtown area, and minibuses were provided to serve these new lots. The short-term effects were substantial. Traffic into the downtown area declined more than 45 percent during the morning peak. No fee was imposed for the afternoon peak, which remained quite congested until 1989, when the afternoon was priced as well. Congestion on the circumferential routes around the downtown area also increased because of the diversion of through traffic.

Many of the former automobile occupants shifted to carpools or traditional bus service (Table B-1). At first, many drivers altered the timing of their trips so that they arrived just before or just after the end of the peak period. Problems in the period before 7:30 a.m. were judged by the authorities to be minor and no actions were taken, but congestion imme-

[6] This discussion about Singapore is largely drawn from Watson and Holland's (1978) evaluation, from Button and Pearman's (1985) description and literature review, and a more recent, extensive review in Annex 1 of Hau's (1992a) paper on congestion charging mechanisms.

[7] Apparently Milan, Italy, has recently instituted a permit system for entering the central area. Although little detail has been published about Milan's system, Orski (1992) reports that automobile trips into the city center have declined by 50 percent and that of this group, 41 percent of motorists shifted to public transit. It appears, however, that this is a system of rationing, not pricing.

TABLE B-1 Mode Used for Work Trips to Restricted Zone from Vehicle-Owning Households Before and After Adoption of Area Pricing (Watson and Holland 1978, Figure 5.1)

	Share of Traffic (%)		
Mode	Before Area Pricing	After Area Pricing	Change
Bus	33	43	+10
Shuttle	—	3	+3
Motorcycle	7	6	−1
Other	4	2	−2
Solo driver	32	20	−12
Carpool	8	19	+11
Taxi passenger	16	7	−9

diately following the pricing period was sufficient to cause them to extend the end of the period from 9:30 to 10:15 a.m., which reduced the congestion. The substantial investments in park-and-ride lots and minibuses did not facilitate a significant mode shift; the shuttles from the lots gained only 3 percent of total trips, which was far less than their capacity. Singapore is currently planning to replace its windshield sticker system with an electronic pricing system (Richards 1992).

An initial evaluation of Singapore's congestion pricing experiment, financed by the World Bank, judged the proposal to have been successful because it met the goal of significantly reducing automobile congestion (Watson and Holland 1978). Button and Pearman (1985) provide a rough calculation of the economic return based on the initial changes in travel time; they estimate a 15 percent return, even when the expenditures on fringe parking lots and feeder buses, which proved far in excess of demand, are included.

Although downtown congestion was reduced substantially during the morning, there have been criticisms. There was some evidence that downtown commercial activity left the area, but Button and Pearman (1985) argue that the evidence is equivocal. There was also a recession under way, which could also explain the relatively minor reduction in businesses in the downtown area. Toh (1977) criticized the plan because it charged too high a fee and argued that, instead of being a traffic management tool, it became a revenue-earner for the government. The diversion of traffic to the ring road also caused considerable delay to travelers on that route, and the time losses to these travelers offset the time savings enjoyed by commuters into the downtown area. Wilson (1988) argues that the number of individuals with increased travel times exceeded those with time savings; his calculations of before-and-after utilities indicate a net loss in social welfare. Wilson (1988) attributes this to the high, flat fee charged.[8] According to Hau (1992a), however, a substantial reduction in the peak-hour fee in 1989, when the program was substantially revised, suggests that this criticism is now less valid. Furthermore, Wilson's analysis does not account for the benefits derived from the uses of the revenues (see footnote 8). The subsequent improvements in the ring road also reduced the time losses to travelers using this facility, which, over time, improved the net benefits of the program (Gomez-Ibanez and Small forthcoming).

The imposition of congestion fees had clear and substantial effects on the traffic entering the downtown area. This benefit has not been disputed, but doubts about the transferability of Singapore's experience to the West have been raised. Singapore has a one-party, quasi-dictatorial government (Morrison 1986; Altshuler 1990). Also, Singapore does not have a separate city government that is distinct from the national government, which "avoids the problem of overlapping jurisdictions between urban and suburban areas that are common, for example, in the United States" (Watson and Holland 1978).

United States

Between 1973 and 1978 the U.S. Department of Transportation (DOT), through the Federal Transit Administration (formerly the Urban Mass Transportation Administration), encouraged U.S. cities to participate in congestion pricing experiments as a means to reduce congestion. Higgins

[8] Wilson attempts to include both the loss in utility from pricing and the gain in utility from using the revenues by recomputing utility using the price and income that prevail after the policy's adoption. However, the assumed utility function does not fully account for income, but rather contains income only in an auxiliary manner as an indicator of value-of-time variations. As explained by Viton (1985) for precisely this kind of situation, the utility function estimated as part of a mode-choice model is necessarily incomplete in accounting for the influence of income, the missing part being that which is independent of mode. Viton shows how this can be taken into account; by not doing so, Wilson in effect omits most of the beneficial effects of the uses of revenue in his comparison of various scenarios. This is why his estimates of utility after the policy are barely affected by whether revenues are returned fully, partially, or not at all.

(1986), one of the Urban Institute staff engaged by DOT to assist the cities that initially responded to Secretary Coleman's invitation, reviewed the individual cases and explained why all efforts foundered. Some of the cities invited to participate refused at the outset but gave different reasons: a lack of congestion (Rochester, New York), fear of controversy (Atlanta, Georgia), and concern about the effect on a struggling downtown area (Baltimore, Maryland). Of those that agreed to participate, in Berkeley, California, the mere possibility of a study resulted in widespread and distorted media coverage, and the mayor and City Council quickly withdrew. After expressing initial interest, the Madison, Wisconsin, City Council declined to participate for a variety of reasons, including concerns about regressivity, appearance of being coercive, and potential harm to the downtown area. The study in Honolulu, Hawaii, progressed the furthest, but policy makers never expressed much support for the concept and, after the first phase of the study was completed, expressed no interest in continuing it.

During this same period, congestion pricing was seriously considered in a major transportation planning exercise involving the California Transportation Commission and the California Department of Transportation. Public hearings on the plan, however, led to considerable criticism from well-organized groups such as the Automobile Club of Southern California (Button 1983). A revision to the plan that reduced the role of pricing as a policy tool did not keep then-Governor Brown from undercutting the entire exercise by eliminating any further budget support for it.

Examination of the efforts to impose higher tolls and congestion tolls on major toll bridges serving San Francisco and New York, and the failure to adopt congestion tolls, reveals some significant problems (Button 1983). In San Francisco, bridge users complained that they did not have alternate routes and therefore that the tolls would be unfair. Others argued that bridge users were not the cause of the downtown congestion. Tolls to restrain use were perceived as inappropriate for facilities paid for by users, and automobile user groups objected. In both San Francisco and New York, downtown merchants, already feeling intense competition from suburban malls, opposed efforts to increase bridge tolls, even though they could have benefitted from reduced congestion. Altshuler (1990) cites similar controversy over a 1986 proposal for areawide pricing in Manhattan. Despite the theoretical benefits of congestion pricing, concerns about fairness, motorist opposition to the concept, and fears of downtown merchants blocked implementation of these plans.

ESTIMATED EFFECTS

Estimates of the level of congestion toll and the likely impact on travel in the United States have been made in several studies (Table B-2).[9] Keeler and Small (1977) estimated long-run optimal congestion tolls for the San Francisco Bay Area. Relying on the earlier work by economists such as Herbert Mohring, Robert Strotz, and William Vickrey, they developed an optimal pricing and investment model. The demand for and peaking characteristics of the San Francisco freeways were calculated from state data on vehicle travel. With these estimates, the model optimized capacity and estimated optimal long-run tolls. The tolls were to vary with the time period and type of route. If these estimated tolls were expressed in 1990 dollars, they would range between $0.05 and $0.36/VMT and would average about $0.15/VMT (Small 1992). Although not estimated by Keeler and Small (1977), the revenue potential of fees this large is quite substantial; if such fees were applied to congested areas of greater Los Angeles, for example, they would raise about $3 billion annually (Small 1992).

To show how short-run optimal tolls would affect different income groups, Small (1983) used a conditional logit model to estimate mode choice for three classes of income groups (high, middle, and low) in response to congestion tolls applied to highly congested facilities. He estimated that to reduce delays for trips involving round-trip delays of 6, 12, and 30 min would require fees of $0.27, $0.98, and $2.20 (in 1972 prices), respectively, per automobile passenger round trip. When the revenues earned were applied to reducing the impact on the low-income group as opposed to being reinvested in highway capacity, all income groups were made better off.

Gomez-Ibanez and Fauth (1980) estimated the effects of areawide licensing and parking fees on travel for the Boston area. Effects were derived from travel demand forecasts (logit model) using a range of elasticities for transit and automobile use. Their results suggested that areawide licensing based on the Singapore model would reduce trips into the center city the most and would also increase transit trips the most. A $1.00 areawide license was estimated to reduce trips to or through the central area by 44 percent.

Recent estimates of congestion tolls were made for the Southern California Association of Governments (Urban Institute and KT Analytics

[9] Many other studies have been conducted abroad to estimate the impact in specific cities.

TABLE B-2 Estimated Effects of Congestion Pricing Policies: Results of Selected U.S. Studies

Author	Type of Pricing	Toll	Estimated Effects
Keeler and Small (1977)	Optimal long-run toll	$0.05 to $0.36/VMT[a] (avg of $0.15/VMT)	Balance of supply and demand
Kraus, Mohring, and Pinfold (1976)	Optimal long-run toll	$0.01 to $0.13/VMT in $1976	Annual welfare losses of at least $1 billion nationwide
Small (1983)	Optimal short run	$0.27, $0.98, $2.20/ passenger round trip (in $1983)	Eliminate delays of 3, 12, and 30 min
Gomez-Ibanez and Fauth (1980)	Area licensing or parking	$1.00/day (in $1980)	Automobile trips to central Boston reduced 40 percent, transit trips up 28 percent, regional VMT reduced 3 to 4 percent
Shoup and Willson (1992a,b)	Cash option in lieu of parking subsidy[b]	NA	Commuter solo driving in Los Angeles area reduced 20 percent, commute VMT reduced 17 percent
Urban Institute and KT Analytics (1991)	All freeways in Los Angeles region	$0.15/VMT	VMT reduced 4 to 6 percent, average commute trip down 10 to 15 min; annual revenues, $2.5 to $2.7 billion
	Area pricing at 20 to 30 major activity centers	$2.00/day	Commute trips reduced 10 percent, trip times reduced 6 to 10 min, annual revenues, $280 million

Kane and DeCorla-Souza (1992)	All routes in a metropolitan area with 1 million commute trips	Average of $0.13/VMT	VMT reduced 2.5 percent; average time savings of 3.6 min/trip; annual revenues, $200 million; hydrocarbons reduced 15 percent
Cameron (1991)	All routes and employee parking	$5.50/vehicle/day	VMT reduced 6.5 percent; fuel consumption reduced 9.5 percent; VOCs and nitrogen oxides reduced 9 to 10 percent
	All routes only	$0.15/VMT or $3.00/day	VMT reduced 5 percent; fuel consumption reduced 9 percent; VOCs and nitrogen oxides reduced 8 percent
Viton (1980)	Optimal short run	$0.15/auto-mile, $0.25/bus-mile, $0.33/truck-mile	Optimizes traffic flow

NOTE: 1 mi = 1.6 km. VMT = vehicle miles of travel. NA = not applicable.

[a] As updated by Small (1992).

[b] Not really pricing but reduction in existing subsidy.

1991). Probable effects of imposing facility pricing on 1280 km (800 mi) of congested urban freeways in the Los Angeles area suggest that tolls of $0.15/VMT would increase travel speeds by 10 to 20 percent, reduce annual vehicle travel by 4 to 6 percent, generate annual revenues of $2.5 to $2.7 billion, and cost about $80 million to $160 million to implement.

Kane and DeCorla-Souza (1992) provided rough estimates of the effects on traffic speed and emissions of a congestion fee averaging $0.13/VMT applied to the principal arterials in a hypothetical urban area with 1 million daily commute trips during the morning and evening peak. They estimated that solo driving would decline by 30 percent, peak-hour vehicle miles of travel would decline by 10.5 percent (total VMT by 2.5 percent), and average commute trips would decline by 3.6 min. Total peak-hour hydrocarbon emissions would decline by 15 percent, resulting in a 3.6 percent decline in total emissions. The tolls would generate about $700,000 per day or about $175 million annually.

Cameron (1991) estimated that a combination of pricing strategies for the Los Angeles area would have substantial effects on congestion and air quality. Charging a peak-hour fee of approximately $3.00 per day per vehicle and significantly increasing employee parking (by $2.25 per day on average) was estimated to reduce VMT by 6.5 percent, mobile source reactive organic gases (the precursors of smog) and nitrogen oxides by 9 to 10 percent, and automobile energy consumption by 9.5 percent. (These estimates are reductions in the travel, energy, and pollution levels that are forecast for the Los Angeles area for 2010 in the absence of any change in policy.) A congestion fee alone of $3.00, which would average about $0.15/VMT, would reduce VMT by 5 percent and reactive organic gases by 8 percent.

Repetto et al. (1992) build on a model developed by Douglass Lee for the U.S. Department of Transportation's 1982 *Final Report on the Federal Highway Cost Allocation Study*. The model estimates the effects of congestion fees on urban roads nationwide. Because some roads are uncongested, no fees would be assessed; on congested urban roads the fees would range from $0.10 to $0.36/VMT. Such fees would reduce peak-period congestion by 22 percent and result in net welfare gains of about $10 billion currently and $21 billion in 1999.

Although Shoup and Wilson (1992a, 1992b) did not explicitly conduct a congestion pricing study, they estimated that employer-subsidized parking is roughly equivalent to about $3.00 per commuter per day, which is about what a $0.15/VMT congestion fee would equal if applied to an average round-trip commute of 20 mi. They applied a logit model to estimate that subsidized parking encourages demand for both solo driving and parking by commuters that is about one-third more than what the demand would be without these subsidies. For the Los Angeles area, they estimated that offering employees the option of receiving a cash allowance in lieu of subsidized parking would reduce commuter solo driving 20 percent and would reduce their VMT by 17 percent. Because peak-period driving accounts for about 40 percent of total daily travel, this would imply an overall VMT reduction of about 7 percent.

The results of the studies reviewed in this section are fairly consistent, even though the methods used and the locales studied vary. The estimates of potential benefits suggest that congestion fees that averaged roughly \$0.15/VMT would reduce peak-period traffic by up to 30 percent in the best case, with reductions in overall travel, energy consumption, and emissions of roughly 5 percent.

REFERENCES

Altshuler, A. 1979. Congestion. In *The Urban Transportation System: Politics and Policy Innovation*, MIT Press, Cambridge, Chap. 9.

Altshuler, A. 1990. Discussion of C. Winston, How Efficient Is Current Infrastructure Spending and Pricing? In *Is There a Shortfall in Public Capital Investment?* Proceedings of a Conference sponsored by the Federal Reserve Bank of Boston, Conference Series No. 34, pp. 206–213.

Button, K. 1983. Road Pricing—An Outsider's View of American Experiences. *Transport Reviews*, Vol. 4, pp. 73–78.

Button, K., and A. Pearman. 1985. Congestion Pricing—Theory and Practice. In *Applied Transport Economics: A Practical Case Studies Approach*, Gordon and Breach Science Publishers, New York and London, Chap. 3.

Downs, A. 1962. The Law of Peak-Hour Expressway Congestion. *Traffic Quarterly*, Vol. 16, pp. 393–409.

Cameron, M. 1991. *Transportation Efficiency: Tackling Southern California's Air Pollution and Congestion*. Environmental Defense Fund, New York City; Regional Institute of Southern California.

Else, P.K. 1986. No Entry for Congestion Taxes? *Transportation Research*, Vol. 20A, No. 2, pp. 99–107.

Giuliano, G., and M. Wachs. 1992. A Comparative Analysis of Regulatory and Market-Based Transportation Demand Management Strategies. In *Papers Presented at the Congestion Pricing Symposium, June 10–12, 1992*, U.S. Department of Transportation, pp. 6-1 to 6-15.

Gomez-Ibanez, J., and K. Small. Forthcoming. *NCHRP Synthesis of Highway Practice: Road Pricing for Congestion Management: A Survey of International Practice*. TRB, National Research Council, Washington, D.C.

Gomez-Ibanez, J. 1992. The Political Economy of Highway Tolls and Congestion Pricing. *Transportation Quarterly*, Vol. 46, No. 3, pp. 343–360. (For summary, see Presentation Summary, In *Exploring the Role of Pricing as a Congestion Management Tool*, Searching for Solutions: A Policy Discussion Series, No. 1, U.S. Department of Transportation, March.)

Gomez-Ibanez, J., and Fauth, G. 1980. Downtown Auto Restraint Policies: The Costs and Benefits for Boston. *Journal of Transportation Economics and Policy*, Vol. 14, pp. 133–153.

Hau, T. 1992a. *Congestion Charging Mechanisms: An Evaluation of Current Practice*. Policy Research Working Paper. WPS 1071. World Bank, Washington, D.C.

Hau, T. 1992b. *An Economic Analysis of Road Pricing: A Diagrammatic Approach*. Policy Research Working Paper. WPS 1070. World Bank, Washington, D.C.

Higgins, T. 1986. Road Pricing Attempts in the United States. *Transportation Research*, Vol. 20A, No. 2, pp. 145–150.

Kane, A., and P. DeCorla-Souza. 1992. Regionwide Toll Pricing: Impacts on Urban Mobility, Environment, and Transportation Financing. In *Papers Presented at the Congestion Pricing Symposium, June 10–12, 1992*, U.S. Department of Transportation, pp. 5-1 to 5-11.

Keeler, T., and K. Small. 1977. Optimal Peak-Load Pricing, Investment, and Service Levels on Urban Expressways. *Journal of Political Economy*, Vol. 85, No. 1, pp. 1–25.

Kraus, M., H. Mohring, and T. Pinfold. 1976. The Welfare Costs of Nonoptimum Pricing and Investment Policies for Freeway Transportation. *American Economic Review*, Vol. 66, No. 4, pp. 532–547.

Knight, F. 1924. Some Fallacies in the Interpretation of Social Cost. *Quarterly Journal of Economics*, pp. 582–606.

Meyer, J., J. Kain, and M. Wohl. 1965. Pricing, Subsidies, Market Structure, and Regulatory Institutions. In *The Urban Transportation Problem*, Harvard University Press, Cambridge, Mass., Chap. 13.

Meyer, J., and J. Gomez-Ibanez. 1981. Traffic Congestion. In *Autos, Transit, and Cities*, Harvard University Press, Cambridge, Mass., Chap. 11.

Mohring, H., and D. Anderson. 1994. *Congestion Pricing for the Twin Cities Areas*. Metropolitan Council of the Twin Cities Metropolitan Area.

Mohring, H., and M. Harwitz. 1962. *Highway Benefits: An Analytical Framework*. Northwestern University Press, Evanston, Ill.

Morrison, S.A. 1986. A Survey of Road Pricing. *Transportation Research*, Vol. 20A, No. 2, pp. 87–97.

Ng, Y. 1977. Towards a Theory of Third-Best. *Public Finance*, Vol 32, No. 1, pp. 1–15.

Orski, K. 1992. Congestion Pricing: Promise and Limitations. *Transportation Quarterly*, Vol. 46, No. 2, April, pp. 157–167.

Pigou, A. 1920. *The Economics of Welfare*. 1st ed.

Repetto, R., et al. 1992. *Green Fees: How a Tax Shift Can Work for the Environment and the Economy*. World Resources Institute, Washington, D.C.

Richards, M. 1992. Road Pricing: International Experience. In *Papers Presented at the Congestion Pricing Symposium, June 10–12, 1992*, U.S. Department of Transportation, pp. 2-1 to 2-10.

Shoup, D., and R. Willson. 1992a. Commuting, Congestion, and Pollution: The Employer-Paid Parking Connection. In *Papers Presented at the Congestion Pricing Symposium, June 10–12, 1992.* U.S. Department of Transportation, pp. 7-1 to 7-21.

Shoup, D., and R. Willson. 1992b. Employer-Paid Parking: The Problem and Proposed Solutions. *Transportation Quarterly*, Vol. 46, No. 2, April, pp. 169–192.

Small, K. 1983. The Incidence of Congestion Tolls on Urban Highways. *Journal of Urban Economics*, Vol. 13, pp. 90–111.

Small, K. 1992. Using the Revenues from Congestion Pricing. *Transportation*, Vol. 19, No. 4, pp. 359–382.

Toh, R. 1977. Road Congestion Pricing: the Singapore Experience. *Malayan Economic Review*, Vol. 22, pp. 52–61.

Transportation Research Board. 1985. *Special Report 209: Highway Capacity Manual.* National Research Council, Washington, D.C.

Urban Institute and KT Analytics. 1991. *Final Report: Congestion Pricing Study.* Southern California Association of Governments, Los Angeles, April.

Vickrey, W. 1959. Statement on the Pricing of Urban Street Use. Joint Committee on Washington Metropolitan Problems, U.S. Congress. *Hearings on the Transportation Plan for the National Capital Region.* Nov.

Viton, R. 1980. Equilibrium Short-Run Marginal-Cost Pricing of a Transport Facility. *Journal of Transport Economics and Policy*, Vol. 14, pp. 185–203.

Walters, A. 1961. The Theory and Measurement of Private and Social Cost of Highway Congestion. *Econometrica*, Vol. 29, No. 4, pp. 676–699.

Watson, P., and E. Holland. 1978. *Relieving Traffic Congestion: The Singapore Area License Scheme.* Working Paper 281. World Bank, Washington, D.C., June. 286 pp.

Wilson, P. 1988. Welfare Effects of Congestion Pricing in Singapore. *Transportation*, Vol. 15, No. 3, pp. 191–210.

Wohl, M., and C. Hendrickson. 1984. Some Practical Pricing Problems. In *Transportation Investment and Pricing Principles*, John Wiley and Sons, New York, Chap. 13.

Zettel, R., and R. Carll. 1964. The Basic Theory of Efficiency Tolls: The Tolled, the Tolled Off, and the Un-Tolled. In *Highway Research Record* 47, HRB, National Research Council, Washington, D.C.

Study Committee Biographical Information

Martin Wachs, *Chairman*, is Professor, Urban Planning Program, at the University of California, Los Angeles, where he is also Director of the Institute of Transportation Studies. Dr. Wachs has taught at the University of Illinois, Chicago Circle, and at Northwestern University. He has authored three books and many articles on urban transportation, public transportation finance and subsidy policy, and the relationships among land use, transportation, and air quality. Dr. Wachs is active in the Transportation Research Board (TRB), the American Planning Association, and other professional organizations.

Jose A. Gomez-Ibanez is Professor, Graduate School of Design, at Harvard University. Before joining the Harvard faculty, Gomez-Ibanez served as Senior Staff Economist for the Council of Economic Advisors. Dr. Gomez-Ibanez is the author and coauthor of many publications on transportation policies and planning. He has served on several TRB committees and is a member of the American Economic Association and the American Planning Association.

Susan E. Hanson is Professor of Geography and Director of the Graduate School of Geography of Clark University. Dr. Hanson has also taught at the State University of New York at Buffalo and at Middlebury College. She has published many papers and articles on the travel patterns of individuals and households in urban areas and on gender issues in local labor markets. Dr. Hanson is an active member and Past President of the Association of American Geographers and is a Fellow of the American Academy of Arts and Sciences.

Steve Heminger is Manager of Legislation and Public Affairs for the Metropolitan Transportation Commission, the metropolitan planning organization for the San Francisco Bay Area. When he joined the study committee, he was Vice President, Transportation, for the Bay Area Council, a regional business-sponsored public policy group. Previously he served as a chief aide to the Chairman of the California State Senate Transportation Committee. Mr. Heminger is also a member of the San Francisco Parking and Traffic Commission.

Joel L. Horowitz is Professor of Economics at the University of Iowa. He has also taught at George Washington University and the Massachusetts Institute of Technology. Dr. Horowitz has written books and articles on transportation systems analysis, econometric analysis of choice behavior, and reduction of adverse environmental impacts of urban transport systems. He is active in TRB, is coeditor or associate editor of several professional journals, and is a member of the Econometric Society and the American Economics Association.

Laura A. Jibben is Executive Director of the Regional Transportation Authority of Northeastern Illinois where she previously was Chief Administrative Officer, among other management positions. She has held several administrative posts in Illinois, including Manager of Toll Services for the Illinois Tollway Department. Ms. Jibben is a member of the National Association of Female Executives and the Women's Transportation Seminar.

Ryuichi Kitamura is Professor of Civil Engineering at the University of Kyoto, Japan. He previously taught at the University of California, Davis. Dr. Kitamura has published extensively on life-cycle and household time-space paths, car ownership and house travel patterns, activity analysis, and effects of aging and motorization on travel behavior. He is active in TRB, the Econometric Society, and the Operations Research Society.

Daniel L. McFadden is Professor of Economics at the University of California, Berkeley. He previously taught at the Massachusetts Institute of Technology and Yale University. Dr. McFadden has published many articles on economics; his work in transportation has been in transportation demand modeling. He has been active in the American Economics Association and served on the Board of Directors of the National Bureau of Economic Research and as the editor of professional journals. He is a past member of the TRB Executive Committee and currently serves as a member of the Commission on Behavioral and Social Sciences and Education. Dr. McFadden is a member of the National Academy of Sciences.

John G. Milliken is with Venable, Baetjer, & Howard of McLean, Virginia. During the course of this study, he was Secretary of Transportation, Commonwealth of Virginia. He is the former managing partner of the Washington, D.C., office of the law firm of Winston and Strawn. Mr. Milliken served for 10 years as an elected member of the Arlington County Board and during that period as a member of the Board of Directors and Past Chairman of the Washington Metropolitan Area Transit Authority, of the Northern Virginia Transportation Commission, and of the Interstate Air Quality Commission for the Washington Metropolitan Area.

Paul E. Peterson is Professor of Government and Director of the Center for American Political Studies at Harvard University. He previously served as the Director of Governmental Studies for the Brookings Institution and has taught at Johns Hopkins University and the University of Chicago. He has authored many articles on intergovernmental relations, education, youth unemployment, and urban policy. Dr. Peterson is a member of the American Political Science Association and the Association of Public Policy and Management.

Mark A. Pisano is the Executive Director of the Southern California Association of Governments. Previously he was the Director of the Environmental Protection Agency's Water Quality Planning Division. He has written several papers on economics and water resources. Mr. Pisano is a member of several regional organizations in Southern California addressing issues such as transportation, housing, and the environment. He is active in the National Association of Regional Councils, the Urban Land Institute, and other professional organizations.

Robert C. Repetto is the Vice President and Senior Economist for the World Resources Institute (WRI). Before joining WRI, Dr. Repetto taught at the Harvard School of Public Health and worked as an economist for the

World Bank. He has authored books and articles on resource economics and conservation, population, and environmental protection. He is a member of the Association of Environmental and Resource Economists, American Economic Association, and the Population Association of America.

William T. Roach is Supervisor, Market Development, Transit Department, King County Metro, located in Seattle, Washington. He previously served as the manager for the Seattle-King County Ridesharing Program. Mr. Roach has been actively involved in developing policies and practices for transportation demand management and land use, and site design guidelines to encourage public transit use, parking management, and special transportation. He is a member of and active in the TRB Transportation Demand Management Task Force and the Association for Commuter Transportation.

Kenneth A. Small is Professor of Economics at the University of California, Irvine. He previously taught at Princeton University and was a Research Associate at the Brookings Institution. Dr. Small has authored books and articles on urban economics, transportation, public finance, and environmental economics, and has contributed to major national studies addressing highway financing policy and traffic congestion. Dr. Small serves on the editorial boards of several professional journals in the fields of urban and transportation studies and is a member of the American Economic Association.

The **Transportation Research Board** is a unit of the National Research Council, which serves the National Academy of Sciences and the National Academy of Engineering. The Board's purpose is to stimulate research concerning the nature and performance of transportation systems, to disseminate the information produced by the research, and to encourage the application of appropriate research findings. The Board's program is carried out by more than 330 committees, task forces, and panels composed of more than 3,900 administrators, engineers, social scientists, attorneys, educators, and others concerned with transportation; they serve without compensation. The program is supported by state transportation and highway departments, the modal administrations of the U.S. Department of Transportation, and other organizations and individuals interested in the development of transportation.

The National Academy of Sciences is a private, nonprofit, self-perpetuating society of distinguished scholars engaged in scientific and engineering research, dedicated to the furtherance of science and technology and to their use for the general welfare. Upon the authority of the charter granted to it by the Congress in 1863, the Academy has a mandate that requires it to advise the federal government on scientific and technical matters. Dr. Bruce M. Alberts is president of the National Academy of Sciences.

The National Academy of Engineering was established in 1964, under the charter of the National Academy of Sciences, as a parallel organization of outstanding engineers. It is autonomous in its administration and in the selection of its members, sharing with the National Academy of Sciences the responsibility for advising the federal government. The National Academy of Engineering also sponsors engineering programs aimed at meeting national needs, encourages education and research, and recognizes the superior achievements of engineers. Dr. Robert M. White is president of the National Academy of Engineering.

The Institute of Medicine was established in 1970 by the National Academy of Sciences to secure the services of eminent members of appropriate professions in the examination of policy matters pertaining to the health of the public. The Institute acts under the responsibility given to the National Academy of Sciences by its congressional charter to be an adviser to the federal government and, upon its own initiative, to identify issues of medical care, research, and education. Dr. Kenneth I. Shine is president of the Institute of Medicine.

The National Research Council was organized by the National Academy of Sciences in 1916 to associate the broad community of science and technology with the Academy's purpose of furthering knowledge and advising the federal government. Functioning in accordance with general policies determined by the Academy, the Council has become the principal operating agency of both the National Academy of Sciences and the National Academy of Engineering in providing services to the government, the public, and the scientific and engineering communities. The Council is administered jointly by both the Academies and the Institute of Medicine. Dr. Bruce M. Alberts and Dr. Robert M. White are chairman and vice chairman, respectively, of the National Research Council.